Informatik aktuell

Herausgeber: W. Brauer
im Auftrag der Gesellschaft für Informatik (GI)

Bernd Reusch (Hrsg.)

Fuzzy Logic

Theorie und Praxis

2. Dortmunder Fuzzy-Tage
Dortmund, 9./10. Juni 1992

Springer-Verlag
Berlin Heidelberg New York
London Paris Tokyo
Hong Kong Barcelona
Budapest

Herausgeber

Bernd Reusch
Universität Dortmund, Lehrstuhl Informatik I
Otto-Hahn-Straße 16, W-4600 Dortmund 50

CR Subject Classification (1992): B.0, I.0, I.2.1, I.2.3, I.2.4, I.2.5, J.2, J.7

ISBN-13:978-3-540-56361-7   e-ISBN-13:978-3-642-78023-3
DOI: 10.1007/978-3-642-78023-3

Dieses Werk ist urheberrechtlich geschützt. Die dadurch begründeten Rechte, insbesondere die der Übersetzung, des Nachdrucks, des Vortrags, der Entnahme von Abbildungen und Tabellen, der Funksendung, der Mikroverfilmung oder der Vervielfältigung auf anderen Wegen und der Speicherung in Datenverarbeitungsanlagen, bleiben, auch bei nur auszugsweiser Verwertung, vorbehalten. Eine Vervielfältigung dieses Werkes oder von Teilen dieses Werkes ist auch im Einzelfall nur in den Grenzen der gesetzlichen Bestimmungen des Urheberrechtsgesetzes der Bundesrepublik Deutschland vom 9. September 1965 in der jeweils geltenden Fassung zulässig. Sie ist grundsätzlich vergütungspflichtig. Zuwiderhandlungen unterliegen den Strafbestimmungen des Urheberrechtsgesetzes.

© Springer-Verlag Berlin Heidelberg 1993

Satz: Reproduktionsfertige Vorlage vom Autor/Herausgeber
33/3140-543210 – Gedruckt auf säurefreiem Papier

Vorwort

Fuzzy Logik fasziniert weltweit Wissenschaftler, Entwickler und Anwender gleichermaßen. Der schnelle kommerzielle bzw. industrielle Einsatz der Fuzzy-Technologie verspricht im zunehmend globalen Wettbewerb erhebliche Vorteile. Allerdings fehlen weithin gezielte Informationen über die Voraussetzungen und Risiken der Entwicklung von Anwendungen.

Die Dortmunder Fuzzy-Tage bieten Entscheidungsträgern und Ingenieuren aus den verschiedensten Bereichen und Branchen Gelegenheit, aktuelle Forschungsergebnisse und Entwicklungen kennenzulernen, Anwendungsfelder für diese äußerst zukunftsträchtige Technologie zu diskutieren und Möglichkeiten für Kooperationen zu erörtern.

Der vorliegende Tagungsband enthält die schriftlichen Ausarbeitungen der Vorträge der 2. Dortmunder Fuzzy-Tage. Diese Veranstaltung fand am 9. und 10. Juni 1992 in Dortmund statt und erfreute sich einer überaus großen Resonanz. Der erste Tag umfaßte mehrere eingeladene Vorträge international renommierter Wissenschaftler, die dem Wissenschaftlichen Beirat der Fuzzy-Initiative Nordrhein-Westfalen angehören. Allen voran ist hier der Nestor der Fuzzy Logik, Prof. Dr. Lotfi A. Zadeh, zu nennen. Diese Vorträge stellen aktuelle Trends und Entwicklungsrichtungen der Fuzzy-Forschung dar.
Der zweite Tag orientierte sich an Erfahrungsberichten aus der industriellen Praxis. Es wurden abgeschlossene, laufende und künftige Entwicklungsvorhaben vorgestellt.

Mein großer Dank gilt an dieser Stelle den Vortragenden sowie den Mitwirkenden an der veranstaltungsbegleitenden Ausstellung für ihr Interesse, ihre Arbeitsergebnisse bzw. Produkte zur Diskussion zu stellen.
Zu besonderem Dank bin ich Herrn Kollegen Zadeh verpflichtet, der trotz seiner vielfältigen Verpflichtungen erneut unserer Bitte nachgekommen ist und das Hauptreferat gehalten hat.

Den Mitgliedern des Organisationsausschusses sei für die Übernahme der nicht unerheblichen organisatorischen Arbeiten gedankt, die mit der Vorbereitung und Begleitung dieser Tagung verbunden waren.

Abschließend möchte ich der Stadtsparkasse Dortmund danken, die der Abendveranstaltung einen angemessenen Rahmen verliehen hat.

Es ist zu wünschen, daß dieses Buch viele aufmerksame und kritische Leser findet und zum Besuch der 3. Dortmunder Fuzzy-Tage anregt.

Dortmund, im Oktober 1992　　　　　　　　　　　　　　　　Bernd Reusch

Inhaltsverzeichnis

Teil I

Modelling with Fuzzy Sets in Fuzzy Control — 3
W. Pedrycz, University of Manitoba, Winnipeg, Canada

The Representation and Use of Uncertainty and Metaknowledge in MILORD, a System for Diagnostic Reasoning — 35
R. L. de Mántaras, Artificial Intelligence Research Institute, Blanes, Spain

Approximate Reasoning and Control with Sparse and/or Inconsistent Fuzzy Rule Bases — 42
L. T. Kósczy, Technical University of Budapest, Budapest, Hungary

Fuzzy Sets and Possibility Theory : Some Applications to Inference and Decision Processes — 66
D. Dubois, H. Prade, Université Paul Sabatier, Toulouse, France

The Calculus of Fuzzy If/Then Rules — 84
L. A. Zadeh, University of California, Berkeley, California

Implementation of Fuzzy Technology — 95
V. A. M. Kwaks; B. W. Grant, OMRON - European Technical Centre, GM's Hertogenbosch, Netherlands

Teil II

MITSUBISHI Fuzzy-Logic im Werkzeug- und Formenbau — 107
O. Keilhofer, MITSUBISHI ELECTRIC EUROPE GmbH, Ratingen

Fuzzy Simulation thermischer Trennverfahren — 112
A. Kraslawski, Lappeenranta University of Technology, Lappeenranta/Finnland; A. Gorak, Universität Dortmund

Anwendung von Fuzzy-Control in der thermischen Verfahrenstechnik — 122
D. Schulz, Ingenieurbüro D. Schulz, Dortmund

Regeln mit 'Fuzzy-Expertensystemen' — 131
H. Knappe, Schröder airporttechnik, Oberschleißheim.

Der Einsatz von Fuzzy-Konzepten
für das operative Produktionsmanagement 143
H.-P. Lipp, MIT GmbH, Aachen;
W. Ringelband, HOESCH Stahl AG, Dortmund

Fertigungsleittechnik mit Fuzzy-Logic 163
U. Schmidt, Simulations-Dienstleistungs-Zentrum, Dortmund

Fuzzy Control Research at Siemens Corporate R & D 179
H. Hellendorn, M. Reinfrank, Siemens AG, München

Online Development Tools for Fuzzy Knowledge-Based Systems
of Higher Order 187
C. von Altrock, B. Krause, INFORM GmbH, Aachen

Entwurf, Simulation und Implementierung von Fuzzy-Logic
in einer integrierten Entwicklungsumgebung 199
M. Hoffmann, TEDAS GmbH, Marburg

Drehzahlregelung eines Gleichstrommotors -
konventionell oder mit Fuzzy-Logic ? 208
Th. Thiel, hema Elektronik GmbH, Aalen

Der Einsatz von Fuzzy-Prozessoren
zur Klassifizierung und Analyse mechanischer Systeme 219
K. Stieger, IDS, München

Die Realisierung einer Fuzzy Logik Hardware - Implementation
am Beispiel des NeuraLogix Fuzzy MicroControllers 227
O. Breiden, UNITRONIC GmbH, Düsseldorf

Touchpad - Fuzzy bewertet Fingerdruck 242
M. Hertzberg, RaPoTronic, Dortmund

Teil 1

MODELLING WITH FUZZY SETS IN FUZZY CONTROL

W. PEDRYCZ
DEPT. OF ELECTRICAL & COMPUTER ENGINEERING
UNIVERSITY OF MANITOBA
WINNIPEG, MB
CANADA   R3T 2N2

Abstract. Fuzzy set-oriented information processing provides a new paradigm in modelling of complex systems. The paper reviews the underlying principle of this type of modelling focusses on its flexibility in handling variable (adjustable) cognitive perpective and distinquishes several general classes of models according to their increasing level of structurial relationships. The detailed identification problems in these classes of models are studied. The vital links between fuzzy control structures and modelling and models emerging there are specified. These are characterized along with several architectures clearly emphasizing the role of fuzzy modelling.

1. Introduction

Fuzzy sets have placed modelling into a new and broader perspective by providing both new methodological and algorithmic tools to cope with complex and ill-defined systems. This area of fuzzy sets has emerged following some pioneering works of Zadeh [34] [35] [36] [37] [49] where the first fundamentals such as fuzzy systems along with their categories and pertaining notation have been introduced. The diversity of currently existing techniques of modelling within the area of fuzzy sets is tremendous.

As far as modelling is concerned, the progress is evident so is a rapid growth of various applications. Nevertheless the methodology of developing and identifying fuzzy models still calls for its better comprehension and more structured, well founded approach.

The main objectives of this study is to reveal the essential facets of fuzzy modelling and point out their role in modelling and control of systems.

First we will study the main aspects of knowledge representation carried out in terms of fuzzy sets. The notion of a fuzzy partition plays a primordial role and provides a flexible realization of the structure suitable for modelling purposes. Different classes of models arranged with respect to a level of structural relationships residing there are reviewed and pertaining identification procedures are examined. Finally the role of fuzzy modelling in formation of control structures and designing control algorithms will be exploited.

The material is structured into sections. Following some preliminaries covered in Section 2 and 3 and dealing with selected aspects of knowledge representation and matching fuzzy quantities we will study the paradigm of fuzzy modelling and, in Section 4, briefly review some representative classes of models (constraints propagation, relational calculus, regression models, etc.). They are discussed in depth in successive sections. The concise examples found in the text are used mainly for illustrative purposes.

2. Knowledge representation conveyed by linguistic labels

The knowledge about the system as well as the perspective from which one is interested to take a look at it is articulated with the aid of linguistic labels. These are generic pieces of knowledge which are deemed by the user as essential in describing and understanding the system. The linguistic labels are represented by fuzzy sets. As demonstrated in [40] they can be also viewed as elastic constraints defined over a universe of discourse and identifying regions with the highest degree of compatibility of elements with the given linguistic term.

Sometimes the linguistic labels are also referred to as information granules. All the information granules defined in the certain space (variable) constitute a frame of cognition of the corresponding variable. [17] [21] The family of fuzzy sets

$$\{A_1, A_2, ..., A_c\}$$

(where $A_i: X \to [0,1]$) constitutes a frame of cognition A if the following two properties are satisfied.
- A "covers" the universe X, namely each element of iG is assigned to at least one granule with a nonzero degree of membership. Formally speaking we obtain

$$\forall_x \exists_i A_i(x) > 0$$

As it will be explained later this property assures that any piece of information defined in X is properly represented by A_i's.
- The elements of A are unimodal fuzzy sets. By stating that we identify several regions of X (one for each A_i) highly compatible with the labels (i.e., with significantly high grades of membership in A_i). The regions defined in this way are characterized by a well-defined semantics.

We distinguish two distinct approaches to developing elements of the frame of cognition.
In the first case the linguistic labels can be specified by studying the problem and recognizing basic relevant information granules which are viewed as necessary in describing and handling it. In this way the subjective evaluation of the membership functions completed by the user of the model becomes a key factor. It is the user who provides relevant membership functions for the variables of the system and therefore creates his own individual cognitive perspective. In this regard some standard methods of membership function estimation apply, see e.g. [32]

The second approach which could be helpful when some records of numerical data are available relies on a suitable utilization of fuzzy clustering techniques. Fuzzy clustering, such as e.g. Fuzzy Isodata [3] enables to discover and conveniently visualize the structure existing in the data set. With the aid of it the numerical data are structured into a number of groups (clusters) according to a predefined proximity measure between data points. The number of clusters is also defined in advance so that they pertain to the linguistic labels constituting the frame of cognition. The algorithm generates grades of membership of the elements of the data set to the given clusters. If necessary these grades can be also converted into an analytical form of final membership functions.

The frame of cognition A can be also referred to as a fuzzy partition of X^1.

We list three essential features of A:

- **specificity** of the frame of cognition. The frame of cognition A' is more specific than A if all the elements of A' are more specific (with specificity defined e.g., in sense of [33]) than the elements of A. Usually a number of elements of A' is greater than the number of the labels in A.

E.g., the frame $A = \{negative, zero, positive\}$ is less specific than the frame

$$A' = \{ Negative\ Large,\ Negative\ Medium,\ Negative\ Small,\\ Zero,\ Positive\ Small,\ Positive\ Medium,\ Positive\ Large\}$$

where now the variable takes on more levels of this linguistic quantification. The partition A' is less general than the previous one. The information granularity of A' is finer than that of A.

- **information hiding** of the frame of cognition refers to each element of A. This feature means that some elements of X are made nondistinguishable by associating them with the same level of membership (usually equal to 1.0). For instance, the fuzzy set A_1 with the membership function defined as

$$A_1(x) = \begin{cases} \text{exponentially increasing over } (-\infty, x') & \text{such that } A_1(-\infty) = 0, A_1(x') = 1 \\ 1, \text{ if } x \in [x', x''] \\ \text{exponentially decreasing over } (x'', \infty) & \text{such that } A_1(x'') = 1, A_1(\infty) = 0 \end{cases}$$

makes all elements from $[x', x'']$ equivalent. By defining this membership function we selectively hide the information about the elements situated within the interval. In other words there is no distinction (at the level of specificity defined by the label A_1) between elements a_1 and a_2, $a_1, a_2 \in X$, as far as both of them are included in the above interval.

Information hiding is completed on purpose so that all following computational processes will not be carried out below this predefined conceptual level. We have already learned that this is an inherent property of the set theory. Fuzzy sets allow to add an extra flexibility to this term by parameterizing it along allowable grades of membership. In other words, λ-cuts are sets completing

---

[1] The fuzzy partition has an additional property: $A_1(x) + A_2(x) + ... + A_c(x) = 1$ which holds for any x; this constraint is automatically satisfied by FUZZY ISODATA but is usually not fulfilled by the first method.

information hiding at this specified level. Particularly for the above trapezoidal fuzzy number used to construct the frame of cognition, their λ-cuts with λ = 1 imply that the information hiding is performed at its highest level.

For the fixed number of the labels, the information hiding can be additionally accomplished by enhancing regions of X associated with the higher grades of membership. For instance, the operation of contrast intensification applied to A [36] [38]

$$\text{Con}(A)(x) = \begin{cases} 2A^2(x) & \text{if } A(x) \in (0, 1/2). \\ 1 - 2(1 - A(x))^2 & \text{otherwise} \end{cases}$$

amplifies "high" values of membership (namely those greater than 0.5) and suppresses those which have already been viewed as insignificant ones.

The information hiding plays an important role e.g., in software engineering [9] although the term coined there has a slightly different technical meaning. The higher generality usually implies the more significant level of information hiding.

Robustness Fuzzy sets constituting the frame A exhibit an interesting property of robustness. Briefly speaking due to smooth transitions in membership of fuzzy sets they allow to tolerate imprecision in the input information to a relatively high extent. Consider the input numerical datum $x \in R$ which due to existing noises and disturbances, is received as x' and as such mapped onto the frame A. This mapping describes levels of activation of $A_1, A_2, ..., A_c \in A$ which become numbers from the unit interval. The noisy version of x induces

$$A_1(x'), A_2(x'), ..., A_c(x')$$

instead of

$$A_1(x), A_2(x), ..., A_c(x)$$

Usually $A_i(x)$ (or $A_i(x')$) is used as an input information of the fuzzy model. The lower the difference between $A_i(x)$ and $A_i(x')$ the higher the robustness of the frame with respect to the input disturbances. Thus the overall sum of absolute differences

$$r(x) = |A_1(x) - A_1(x')| + |A_2(x) - A_2(x')| + ... + |A_c(x) - A_c(x')| \quad (1)$$

can be viewed as a suitable indicator of the robustness property of A. For $r \approx 0$ we can talk about tolerance of incorrect (false) information. It should be stressed that the above measure of fault tolerance[2] becomes a function of x. Therefore it could vary significantly from point to point of

---

[2] One should note that the term of fault-tolerance used here refers to a s-called "faulty" information as opposed to fault-tolerance pertaining to faults in a structure, cf. [11].

X. The overall robustness measure can be defined by completing standard averaging over the universe of discourse.

$$r = \int r(x) \, dx \qquad (2)$$

(We assume that this integral taken over X does make sense).

It is also obvious that (2) can attain low values since $A_i(x)$ and $A_i(x')$ can generate relatively similar values of membership (for relatively low values of disturbances). In the case of sets, $A_i(x)$ and $A_i(x')$ could have distinct values of membership even for some very close values of x and x', say

$$x \approx x' \quad \text{but} \quad A_i(x) = 0 \text{ and } A_i(x') = 1$$

Evidently if both x and x' are included in the support of A_i (or both are excluded from it) then the value of r at a certain element of X is equal to zero. The error attains maximal values at the borders of the set.

Example 1. We will study the frame composed of three linguistic labels defined in the interval [1,5]. Their membership functions are Gaussian-like described by modal values m_i and spread values δ_i,

$$A_i(x) = \exp[-(x - m_i)^2/\delta_i]$$

See Fig. 1

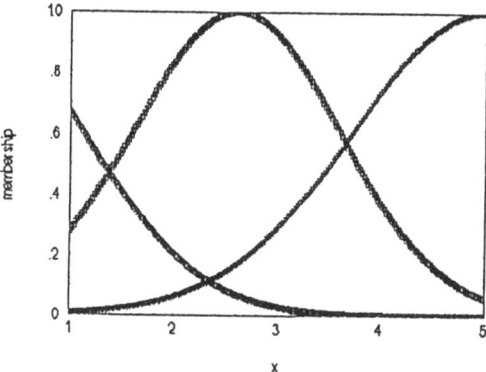

Fig. 1 Linguistic labels of A

The noise is modelled as a random variable of the Gaussian distribution with a zero mean value and a standard deviation σ, $N(0,\sigma)$. The standard deviation σ refers to the level of disturbances. The frame is exposed to inputs (x's) uniformly distributed over the space and affected by the Gaussian noise. The resulting values of the robustness measure are computed for each discrete point of the interval. The experiment is repeated (averaged) 20 times to visualize a general tendency in the values of r(x). The obtained results (for several values of the standard deviation of the noise, σ = 0.25, 1.0, 3.0) are summarized in Fig. 2

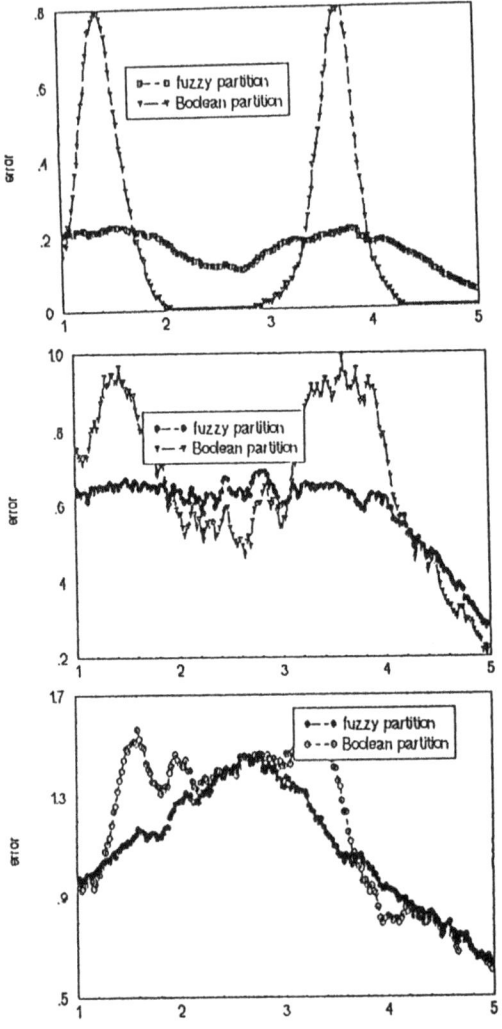

Fig. 2 r(x) for fuzzy partition and the Boolean version (a) σ=0.25 (b) σ=1.0 (c) σ=3.0

It is worthwhile to compare the performance of r(x) generated by the Boolean counterpart of the frame A. The induced Boolean frame consists of these sets. They are formed by determining points of intersection of A_1 and A_2 as well as A_2 and A_3, say x_1 and x_2. Then we distinguish sets as three disjoint intervals $[1, x_1]$, (x_1, x_2), $[x_2, 5]$. The results are summarized again in Fig. 2.

In comparison to those obtained for the fuzzy partition several meaningful differences are noticeable. Firstly, the local values of r(x) are significantly higher than these for the fuzzy partition. Secondly, the maximal discrepancies occur at the borders of the sets where there is no "stabilizing" effect of the smooth transitions of membership functions. Overall, the averaged robustness is higher for the fuzzy partition. The histograms of the values of "r" summarize the difference in a transparent form. The distribution of the error is more concentrated for the fuzzy partition; the Boolean partition generates both zero as well as high values of "r".

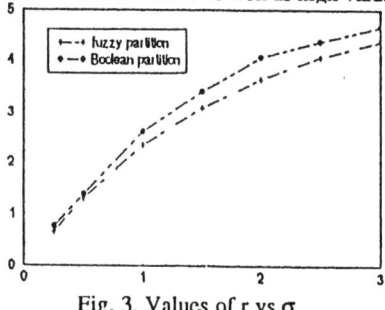

Fig. 3. Values of r vs.σ

It is worth indicating that the fuzzy partition leads to a homogeneous form of information. To elaborate a bit on this issue let us note that an input information can be uniquely expressed in terms of linguistic labels. The converse is not true; knowing the representative of the input information at the level of the information granules one cannot reconstruct on in a unique manner. This phenomenon is due to the level of generality introduced by fuzzy sets yielding a set-to-point mapping. The scheme of transformation can be portrayed schematically as follows:

$$\text{input information} \rightarrow \text{fuzzy partition} \rightarrow x$$

where the input information could be given in a numerical, interval, or fuzzy set format. The derived output x expresses this input information by providing degrees of matching (activation) of the elements in the fuzzy partition. Thus x becomes associated with the fuzzy partition and its semantics directly reflects the semantics of the linguistic labels. In a simple example where the input information is given precisely as a single numerical quantity x_0, the vector $x \in [0,1]^c$ can be developed by taking the values of the possibility measure [7] [39] of x_0 expressed with respect to $A_1, A_2, ..., A_c$, namely

$$x = [A_1(x_0) \ A_2(x_0) \ ... \ A_c(x_0)]$$

For interval-valued information $[x_1, x_2]$ the resulting vector x is expressed as

$$x = [\sup_{x \in [x_1,x_2]} A_1(x) \quad \sup_{x \in [x_1,x_2]} A_2(x) \quad ... \quad \sup_{x \in [x_1,x_2]} A_c(x)]$$

This is reflected by higher values of the entries of x. Obviously

$$\sup_{x \in [x_1,x_2]} A_i(x) \geq A_i(x_o)$$

for $x_o \in [x_1, x_2]$

Of course the interval-valued information is less precise than a single numerical quantity. The limit case (all elements of the space) causes that all the entries of x are equal to 1 (assuming that the labels are normal fuzzy sets).

3. Expressing equality between fuzzy sets: equality index and inverse problem

We will briefly report on an operation on fuzzy sets which is not a standard one and will be used in further studies. It is used to carry out comparisons between fuzzy sets and is named an equality index. The associated concepts are equality intervals. A detailed discussion is contained in [18]. Let us study two fuzzy sets A,B defined in the same universe of discourse. The equality $A \equiv B$ is defined pointwise as

$$A \equiv B = [a_1 \equiv b_1 \quad a_2 \equiv b_2 \quad ... \quad a_n \equiv b_n]$$

where

$$a_i \equiv b_i = \frac{1}{2}\left[(a_i \; \varphi \; b_i) \wedge (b_i \; \varphi \; a_i) + (\bar{a}_i \; \varphi \; \bar{b}_i)(\bar{b}_i \; \varphi \; \bar{a}_i)\right] \quad (3)$$

and the φ-operator is induced by an associated t-norm,

$$a_i \; \varphi \; b_i = \sup\{c \in [0,1] \mid a_i \; t \; c \leq b_i\}$$

Identifying the φ-operator with a certain multivalued implication operation, the above definition has an interesting set-theoretic interpretation. The first part of the expression reads as a conjunction of the two inclusion conditions (notice that the implication operation is equivalent to a multivalued inclusion relation),

$$(a_i \subset b_i) \wedge (b_i \subset a_i)$$

The second part of (3) expresses the same property holding between the complements of a_i and b_i.
In a so-called inverse problem associated with the equality index one looks for B assuming that

A and the equality vector $\gamma = A \equiv B$ are specified. More formally we will write the problem down as the set of inequalities

$$a_i \equiv b_i \geq \gamma_i \quad i = 1,2,\ldots,n \tag{4}$$

Which has to be solved with respect to b_i

It has been proven [18] that the inequalities always have a nonempty set of solutions. Furthermore, the solution set

$$B_i = \{b_i \mid a_i \equiv b_i \geq \gamma_i\}$$

constitutes an interval in [0,1] for each element of the element of the universe. The uniqueness of the solution is assured only for $\gamma_i = 1$. The lower the value of γ, the broader the interval. This emphasizes that lower γ's imply an increasing uncertainty tied with the solutions of the inverse problem - no unique solution exists and our ability to identify this solution diminishes. Since the interval expands, another interesting look at the inverse problem could underline the nature of the solution - it becomes an interval-valued fuzzy set [18] [20] with equally possible grades of membership located within B_i.

4. System modelling with fuzzy sets: The paradigm of fuzzy modelling

The basic idea of fuzzy models and fuzzy modelling is to model system at a level of linguistic labels. One should underline that the models of this class are not used to represent relationships between variables at a numerical, pointwise level. The role of fuzzy modelling is to look at the system from a "distance" by accepting a proper cognitive perspective. The fuzzy partition constructed for each variable can be adjusted separately to meet requirements of the modelling task. Some details can be then selectively hidden and won't increase unnecessary computational burden of model building[3]. If more details are required one can formulate another more detailed fuzzy partition and re-design the fuzzy model. In a limited case a purely pointwise (numerical) model can emerge. The fuzzy models developed in this way concur with the principle of incompatibility formulated by Zadeh [36]; a formulation similar in its essence can be found in Puccia and Levins, [23]. The principle states that any model building calls for a rational trade-off between significance (relevance) and precision achievable within the model. It should be stressed that one should sacrifice (to a certain degree) precision to reach an acceptable level of generality. Overall, the properties of the frame of cognition applied to modelling are inherited and become in-built into the fuzzy models. For instance, by increasing robustness of the frame A one implies higher robustness of the fuzzy model. The overall structure of fuzzy modelling as combining, processing, forming, and interpreting data is schematically shown in Fig.4.

---

[3] In a so-called qualitative modelling all relationships are defined between symbolic quantities, cf. [6]; the extenstion including linguistic variables is reported in [1].

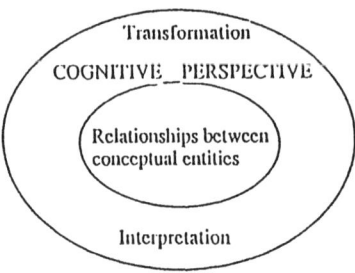

Fig. 4 Combining processing and interpreting data in fuzzy modelling

Through the fuzzy partition one can easily process heterogeneous pieces of data and provide interpretation of the results of modelling in the same linguistic categories of the fuzzy partition. The core part of the identification activity will be therefore concentrated on expressing links between linguistic labels. In sequel we will be concerned with constructing these formal relationships. The discussion to follow will distinquish several important classes of the fuzzy models with significant variations in their structural characteristics.

4. Classes of fuzzy models

4.1 Introductory remarks

In this section we will review several classes of fuzzy models, highlight their essentials as well as comment on their level of structurability. Most of them will be studied in detail in later sections and identification algorithms will be provided. We will start with the class of models of constraint propagation with the lowest level of imposed structural relationships. Then we will proceed with more structured classes of models such as relational or neural network-oriented and those based on highly structured fuzzy arithmetic completed for system's variables.

4.1.1 Constraint Propagation

The model based on the principle of constraint propagation does not assume any particular structure which is conveyed by the data set. The frame of cognition is formed for all input variables and all the fuzzy sets involved there are propagated through the data. For some combinations of the linquistic labels the data support the input fuzzy sets, and generate a corresponding fuzzy set in the output space. In some others, the input combination does not get enough numerical support becomes filtered out by the data and does not appear in the model. In general, the model is formed by identifying combinations of the linguistic labels for input and output variables which are sufficiently supported by the numerical data.

4.1.2 Relational calculus and relational models

The models based on relational calculus express links between linguistic labels in terms of relations. In other words, the statements about system's variables are expressed as

"there is a relationship between A and B" (A and B are related)

or more formally

$$A R B$$

where A and B are linguistic labels in two universes of discourse in which the variables are defined. The relation R in the above expression is a fuzzy one and it specifies a degree to which the objects A and B are related (interact). The identification algorithms as well as different modes of utilization of this class of models makes an extensive use of fuzzy relational equations and inequalities, see for instance [5] [15] [16]. This area is theoretically well-developed with numerous computing algorithms available there [19] [20]. The models of this class usually involve single level relational structures (articulated by a single relation holding between A and B). The differences between the models lie with respect to composition operators playing a key role.

4.1.3 Fuzzy neural networks

Fuzzy neural networks are used as distributed models of systems and are perceived as collections of logical (i.e., driven by logical operations) neurons. This class of models is characterized by significant adaptability (learning) properties and additional logic-specific representation capabilities associated with the constructs. Fuzzy neural networks originated as a natural extension of the fuzzy relational models, which, in this context are simple two-layer neural networks. The studies on these models have been initiated in [16] [19]; refer also to some existing application-oriented studies [22].

4.1.4 Fuzzy regression models

Fuzzy linear regression models introduced in [30] [31] constitute another approach toward fuzzy identification and modelling. The underlying idea is that the model is described as a linear function of input variables with its parameters viewed as fuzzy numbers,

$$Y = A_1 x_1 + A_2 x_2 + \ldots A_n x_n$$

Where $x_1, x_2, \ldots, x_n \in \mathbf{R}$ are independent variables of the model and $A_1, A_2, \ldots, A_n$ are usually triangular (or trapezoidal) fuzzy numbers. Each of these parameters A_j is a piecewise linear membership function of a, say

$$A_j(a) = \begin{cases} 1 - \dfrac{|\alpha_j - a|}{c_j} & \text{if } \alpha_j - c_j \leq a \leq \alpha_j + c_j \\ 0 & \text{otherwise} \end{cases}$$

$a \in R$

The parameter of the membership function α_j is known as a modal value of the fuzzy number A_j. The modal value describes the most possible value of the parameter. The second parameter, c_j, denotes a spread of the fuzzy number which determines a distribution of the number (its ambiguity) around the modal value. The spread characterizes the precision of the parameter. In fact the values of c_j's are used in order to absorb deficiencies of the constructed model, as well as to incorporate ("neutralize") noise existing in the data set. The higher the value c_j, the greater uncertainty.

Two interesting classes of fuzzy numbers are worth studying. Firstly, the triangular number A can be compactly described by three real numbers arranged in an increasing order, namely

$$A = (m\_lower, m, m\_upper)$$

where the support of A is constrained by the limit values m_lower and m_upper. The modal value is denoted by "m". The membership function increases linearly over the arguments ranging from m_lower to m, A(m_lower) = 0, A(m) = 1 and decreases in the same linear manner achieving 0 at the upper limit, m_upper. The class of trapezoidal fuzzy numbers admits a certain range of arguments with the grade of membership equal to 1. The four-number notation of any trapesoidal fuzzy number reads as

$$A = (m\_lower, m_1, m_2, m\_upper)$$

where $A(x) = 1$ for all $x \in [m_1, m_2]$. Again for $x \in [m\_lower, m_1]$ the membership function grows linearly and afterwards decreases linearly over $[m_2, m\_upper]$.

The fuzzy regression models are characterized by a higher level of structural linkage in comparison to those previously discussed since they heavily utilize the linear form of input-output relationships (even though they are re-expressed in terms of fuzzy numbers).

The identification algorithms are well defined and computationally efficient. In fact they reduce to standard schemes of Linear Programming. The models of this class can be also extended to nonlinear ones with respect to input variables. Nevertheless, this extension preserves linearity of the model with respect to its parameters.

4.1.5 Local regression models

The idea behind this class of fuzzy models introduced in [29] is to replace a single and a universal model by a series of "local" models. These are usually easier to construct and verify.

They are valid in certain local and constrained regions of input variables $x \in R^n$. The global model is formulated with the aid of conditional statements including the local models constructed so far,

- if $x \in \Omega_1$ then $y = \psi_1(x, a_1)$
- if $x \in \Omega_2$ then $y = \psi_2(x, a_2)$
  $\vdots$
- if $x \in \Omega_p$ then $y = \psi_p(x, a_p)$

The condition parts of the above rules specify regions Ω_j of the input variables, namely $\Omega_j \subset R \times R \times \ldots \times R$. The conclusion parts consist of linear or nonlinear relationships ($\psi_j(x, a_j)$) which are relevant in the region given in the corresponding condition part and do not apply to outside the associated regions. To make the model complete, the family $\Omega_1, \Omega_2, \ldots, \Omega_p$ should form a Boolean partition of the n-fold Cartesian product of R. The replacement of the Boolean regions by their fuzzy set based versions enables us to avoid eventual discontinuities occurring in the output values of the model. In a certain neighbourhood of the regions Ω_i it could happen that $|\psi_i(x, a_i) - \psi_j(x', a_j)| > \delta$ even for $x \approx x'$. To alleviate this evident drawback we replace Ω_i's by a fuzzy partition. The fuzzy partition introduces a necessary continuity between the local models (ψ_i). In case all of the "local" models are linear ($\psi_i(x, a_i) = x^T a_i$) the method suggested in [20] made it possible to determine their parameters (a_i) through the use of standard regression techniques. Once the presentation of the general classes of fuzzy models has been completed we will study them in a more detailed format.

5. Constraint propagation

The main idea of the method of constraint propagation is to elicit qualitative dependences between predefined linguistic labels for independent variables of the system and the labels for output variables of the system. Once the linguistic labels have been defined they are viewed as constraints imposed on data and "propagated" along the data available for identification purposes. The propagation process will be described later. The data points cause that some combinations of the linguistic labels are supported by them and generate linguistic labels for the output variables. Some others (usually a significant portion of them) do not gain any significant support and are eliminated from the model description.

As a concise illustration let us consider a system with two input variables and a single output. For the first input variable we assume that the linguistic label is given by a fuzzy set A; for the second variable we will denote the label by B. The data set used for identification purposes is a collection of N-triples of data: (input variable$_1$, input variable$_2$, output). We accept nonfuzzy data i.e., pointwise values (x_k, z_k, y_k), $k = 1,2,\ldots,N$. Definitely one can admit a generalized character of the data this could lead to a slight modification of the discussed algorithm.

Let us construct α-cuts of fuzzy sets A and B, A_α and B_α. They play a role of constraints imposed on the structure of the system. At the same time they direct a focussed search through the data. A_α and B_α "covers" some portion of the data viz. all x_k and y_k such that $x_k \in A_\alpha$, $y_k \in B_\alpha$.

The induced subsets of y_k denoted by $(\{y_k\}|A)_\alpha$ will be formally put down as

$$(\{y_k\}|A)_\alpha = \{y_k \mid x_k \in A_\alpha\}$$

Similarly one gets

$$(\{y_k\}|B)_\alpha = \{y_k \mid z_k \in B_\alpha\}$$

By intersecting the above sets we jointly propagate the constraints A and B (more exactly their α-cuts) through the data set

$$(\{y_k\}|A \& B)_\alpha = \{y_k \mid x_k \in A_\alpha\} \cap \{y_k \mid z_k \in B_\alpha\}$$

The results of this propagation can be treated as a fuzzy set C defined for the output variable. Symbolically one writes it down as

$$A,B \text{ - propagation through - C}$$
$$\text{the data}$$

The above expression describes α-cuts of C which, due to the representation theorem [36], is reconstructed by taking a union of C_α's.

$$C = \bigcup_{\alpha \in [0,\pi]} \alpha \, C_\alpha$$

where π stands for the highest value of the membership function of C (height of C)

$$\pi = \text{hgt}(C)$$

which corresponds to the highest value of α yielding a nonempty intersection,

$$\{y_k \mid x_k \in A_\alpha\} \cap \{y_k \mid z_k \in B_\alpha\} \neq \phi \quad \text{for } \alpha \leq \pi$$

$$\{y_k \mid x_k \in A_\alpha\} \cap \{y_k \mid z_k \in B_\alpha\} = \phi \quad \text{for } \alpha > \pi$$

The height of the generated fuzzy set π can be used as a performance measure expressing a coincidence of the labels with the data. This type of possibility-based criterion can be accompanied by the probabilistic type of criterion by which one expresses the following probability

$$p(\alpha) = \frac{\text{number of data included in } C_\alpha}{\text{total number of data (N)}}$$

Since $C_{\alpha_1} \supset C_{\alpha_2}$ for $\alpha_1 > \alpha_2$ then $p(\alpha)$ is a nonincreasing function of α. Again the constraints A and B with a low value of the probability $p(\alpha)$ should be discarded as a meaningless combination of the labels.

In case of "n" input variables and several linguistic labels the procedure of constraint propagation should be repeated separately for each of them. The combinations with the highest value of π will be used as legitimate components (descriptions) of the model.

For "n" variables with "m" linguistic labels (fuzzy sets) we have m^n possible combinations. Usually a significant portion of them is ruled out. The choice of the labels is based on the analysis of the derived fuzzy sets of the output variable and include overlap between them as well as their heights or associated probabilities. The trade-off exists between the overlap of these fuzzy sets (which calls for more specific labels for the input variable) and a lack of a complete coverage ("explanation") of all the data provided by the constraints. The last phenomenon suggests the use of coarser constraints (of lower granularity) with a higher overlap.

We will illustrate the performance of the method discussing the following example.

Example 2: In this example we will construct a fuzzy model for a set of data coming from [4]. The data set consists of results of 31 observations of a chemical process in which tar content of an output was related to the temperature at which the process was run and to the speed of a rotor, see Fig. 5

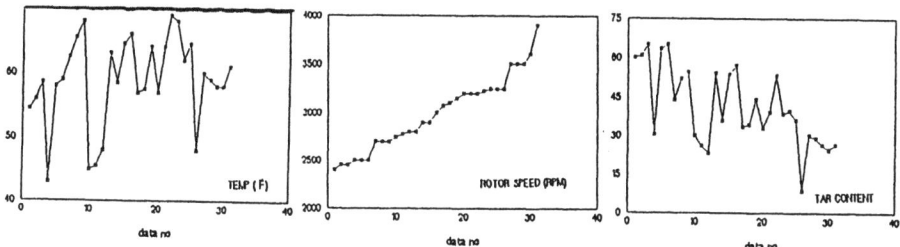

Fig. 5 Experimental data of a two-input single output process

To obtain a qualitative description of the process we defined a relevant frame of cognition for each of the input variables (here the temperature and the speed). Numerical experiments include a different number of labels for the input variables, namely 2, 4, and 6. The labels are specified as triangular or trapezoidal fuzzy numbers.

The choice of the fuzzy sets can be guided by an empirical probability distribution of variables of the respective values. By studying their (usually multimodal) histograms, one can decide on suitable linguistic labels, see Fig.6.

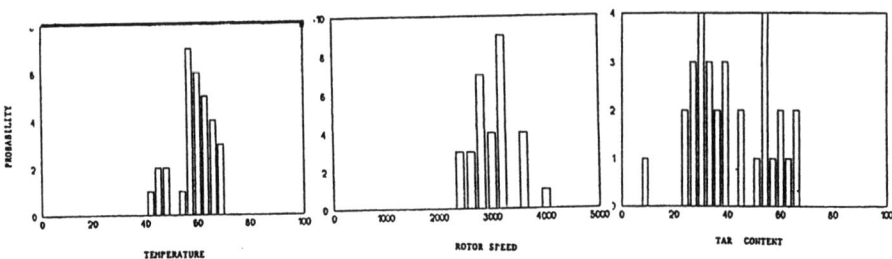

Fig. 6 Histograms of values of input and output variables

Of course, this type of guideline constitutes one of the feasible options.

The resulting fuzzy sets of the output variable are characterized both in terms of a total probability assigned to them as well as their height. Furthermore their maximal α-cuts are summarized. Table 1 combines all the derived results (some entries of the tables are left empty since the data did not support any propagation of the corresponding input constraints yielding low values of π).

temperature
$A_1 = (35, 40, 46)$
$A_2 = (45, 50, 55)$
$A_3 = (50, 56, 62)$
$A_4 = (59, 70, 75)$

rotor speed (all numbers to be multiplied by 10^3)
$B_1 = (2.4, 2.5, 2.7)$
$B_2 = (2.6, 2.8, 2.9)$
$B_3 = (2.8, 3.0, 3.6)$
$B_4 = (3.4, 4.0, 4.8)$

	B_1	B_2	B_3	B_4
A_1	30.5 $\pi = 0.48$ $p = 0.032$	30.0 $\pi = 0.15$ $p - 0.065$	—	—
A_2	—	23.0 $\pi = 0.6$ $p = 0.032$	8.5 $\pi = 0.5$ $p = 0.032$	—
A_3	63.5 $\pi = 0.65$ $p = 0.032$	—	33.5 $\pi = 0.83$ $p = 0.032$	24.5 $\pi = 0.33$ $p = 0.032$
A_4	—	[52.0, 54.5] $\pi = 0.48$ $p = 0.161$	53.0 $\pi = 0.65$ $p = 0.032$	26.5 $\pi = 0.175$ $p = 0.065$

Tab. 1 Results of modelling for some selected levels of α

5. Fuzzy relational structures

The models of this class capture dependencies between linguistic labels and express them as fuzzy relations. Fuzzy sets, fuzzy relations, and the calculus of these objects are put together into a form of fuzzy relational equations. Originally developed outside fuzzy sets as a branch of relational calculus and applied to problems of operations research, cf. [24] afterwards the equations were reformulated and generalized in [25]. Some interesting links with multivalued logic have been underlined in [12]. Since then many theoretical results have been obtained and the methodology of their use followed by a series of specific applications has been formulated.

We will be concerned with discrete versions of fuzzy relational equations namely the equations defined in finite universes of discourse. The fuzzy sets x and y include levels of activation of the frames in which input and output variables are expressed, thus $x \in [0,1]^n$, $y \in [0,1]^m$. The general statement

$$x \text{ and } y \text{ are related } (R)$$

can be translated in many ways. The main classes of relational structures (and subsequently fuzzy relational equations) are summarized below. The fuzzy set y results from x and R as

$$y = x \circ R$$

$R \in [0,1]^{n \times m}$. The composition operator "o" involves triangular norms (t- and s-norms)

$$y_j = \mathop{S}_{i=1}^{n} [x_i \, t \, r_{ij}]$$

$j = 1, 2, ..., m$.

Depending on the combination of the triangular norms (selected as the maximum and the minimum operator) one has a well-known max-min composition used in most of existing constructs in fuzzy sets. The next step of generalization is to allow any continuous t-norms in place of the minimum operation.

The relational structure can be treated as a relational equation with two generic problems formulated accordingly
  i) x and y are given, determine R
  ii) y and R are given, determine x

For the max-min and max-t composition there are analytical solutions available. Furthermore the nonempty family of solutions contains more than a single element. The extremal solutions (maximal ones) are determined with the use of the residuation operation (φ-operator [15]) associated with the t-norm standing in the equation, refer also to Section 3.

The fundamental results are then concisely summarized:

for (i): if the family of solutions $R \neq \phi$ then the maximal element of R, $\widehat{R} = \max R$ is determined through the φ-operation applied pointwise to $\mathbf{x}$ and $\mathbf{y}$.

$$\widehat{R} = \mathbf{x} \, \varphi \, \mathbf{y}$$

i.e.

$$\widehat{r}_{ij} = x_i \, \varphi \, y_j$$

for (ii): assuming that the family of solution X is nonempty, its maximal element is calculated

$$\widehat{\mathbf{x}} = R \, \varphi \, \mathbf{y}$$

or

$$\widehat{x}_i = \min_{j=1,2,\ldots m} (r_{ij} \, \varphi \, y_j)$$

For the general case (s-t composition) one cannot develop analytical solutions to the relational equations. There are different interpretations of the fuzzy relational equation with the max t operator. The one is that in which for a fixed j, $j = 1,2,\ldots,m$ we can interpret y_j as a height of intersection of $\mathbf{x}$ and $\mathbf{r}_j$ (the j-th column of the fuzzy relation R).

$$y_j = \text{hgt}(\mathbf{x} \cap \mathbf{r}_j)$$

The dual class of fuzzy relational equations involve t-s composition (dual to the previous composition). Now

$$\mathbf{y} = \mathbf{x} \, \Delta \, R$$

$$y_j = \mathop{T}_{i=1}^{n} (x_i \, s \, r_{ij})$$

$j = 1,2,\ldots,m$.

Again the two more specialized families of composition include min-max and min-s aggregation. The analytical solutions are available only for these families. The character of the results is dual to those reported previously. Briefly, the obtained solutions are minimal in the family of solutions:

- for (i)

$$\check{R} = \mathbf{x} \, \beta \, \mathbf{y} \qquad r_{ij} = x_i \, \beta \, y_j$$

- for (ii)

$$\check{\mathbf{x}} = R \, \beta \, \mathbf{y} \qquad \check{x}_i = \max_{j} \, [r_{ij} \, \beta \, y_j]$$

where the definition of β-operator uses the s-norm standing in the equation

$$a \beta b = \inf\{c \in [0,1] \mid a \, s \, c \geq b\}$$

The adjoint fuzzy relational equations are defined with respect to the direct and dual equations. For the direct equations one has

viz.
$$y = x \, \varphi \, R$$

$$y_j = \max_i [x_i \, \beta \, r_{ij}]$$

The analytical solutions are extremal in the entire family of solutions and are constructed with the aid of t- and s-norms. [13] [7] The main results summarizing both relevant formulas and the character of solutions are shown below.

type of equation	generic equation	solution with respect to X	solutions with respect to R	character of solutions
direct	$y = x \circ R$	$x_i = \min_j (r_{ij} \, \varphi \, y_j)$	$r_{ij} = x_i \, \varphi \, y_j$	maximal
dual	$y = x \, \Delta \, R$	$x_i = \max_j (r_{ij} \, \beta \, y_j)$	$r_{ij} = x_i \, \beta \, y_j$	minimal
adjoint to direct	$y = x \, \varphi \, R$	$x_i = \min_j (r_{ij} \, \varphi \, y_j)$	$r_{ij} = x_i \, t \, y_i$ *	minimal
adjoint to dual	$y = x \, \beta \, R$	$x_i = \max_j (r_{ij} \, \beta \, y_j)$	$r_{ij} = x_i \, s \, y_i$ *	maximal

* t- and s-norm

Solutions to fuzzy relational equations and their characteristics

Overall, two facts are worth noting:
- for the nonempty solution set the analytical methods provide extremal elements in the space of solutions (maximal or minimal ones). Building some other solutions requires auxiliary search strastegies.
- the solutions to systems of equations are again based exclusively on these extremal solutions for individual equations. This again makes them extremal to the entire family of solutions.

The theory holds under a strong assumption that the relevant equation(s) is solvable. If this is not fulfilled one should look for approximate solutions since, due to their extremal nature, the previous findings should be treated with caution and could not constitute good candidates even for approximate solutions. For more detailed discussion the reader is referred to [16] [20] [5].

7. Fuzzy neural networks

In this section we will take another look at more structured family of fuzzy models by studying fuzzy neural networks. By this class of networks we mean distributed and parallel computing structures heavily employing logical operations existing in the theory of fuzzy sets. As opposed to standard neural networks the networks emerging within this framework are usually heterogeneous i.e., they consist of neurons of a different conceptual and numerical nature. When put together they exhibit diverse functional characteristics and play quite distinct roles in the network. First we will discuss basic models of neurons (aggregative and reference ones) and afterwards concentrate on logical processor constituting a generic architecture of the fuzzy neural networks. In sequel learning algorithms are studied. Representation and generalization aspects are considered as well.

7.1 Aggregative and matching functions of logical neurons

The logical neurons aggregate input signals $x_1, x_2, \ldots, x_n \in [0,1]$ in a logical form. Two basic logical connectives AND and OR give rise to so-called AND and OR neurons. The AND neuron AND's the input signals

$$y = x_1 \text{ AND } x_2 \text{ AND } \ldots \text{ AND } x_n \tag{6}$$

For the OR neuron we obtain

$$y = x_1 \text{ OR } x_2 \text{ OR } \ldots \text{ OR } x_n \tag{7}$$

Both the AND and OR connectives are represented as triangular norms (t- and s-norms). Further extensions of the above formulas include weights (connections) associated with inputs so that an influence of x_i's on the output y can be conveniently accommodated. In this way (6) and (7) are translated into the following formulas

AND neuron

$$y = (x_1 \text{ OR } w_1) \text{ AND } (x_2 \text{ OR } w_2) \text{ AND } \ldots \text{ AND } (x_n \text{ OR } w_n) \tag{8}$$

$w_i \in [0,1]$, $i = 1,2,\ldots,n$

OR neuron

$$y = (x_1 \text{ AND } w_1) \text{ OR } (x_2 \text{ AND } w_2) \text{ OR } \ldots \text{ OR } (x_n \text{ AND } w_n) \tag{9}$$

The weights are used to enhance or eliminate an influence of x_i's on the output y. For the AND neuron we conclude:
- the lower the value of w_i the more evident influence of x_i on y
- higher values of w_i enhance the importance of x_i.

In limit cases ($w_i = 0$ for the AND neuron and $w_i = 1$ for the second one taken for all variables) the formulas (8) - (9) reduce to (6) and (7), respectively.

The AND (OR) neuron can be enhanced functionally in two different ways:
- the complemented input signals, $\bar{x}_i = 1 - x_i$, allow to realize inhibitory performance of the neuron still preserving the unit interval as a suitable range of coding. By choosing appropriate values of connections the neuron can easily exhibit inhibitory and excitatory characteristics. The simple example illustrating this deals with a neuron with two inputs x and $\bar{x}$ employing standard max and min operations. Its characteristics depending on the values of connections w_1 and w_2 are visualized in Fig.7

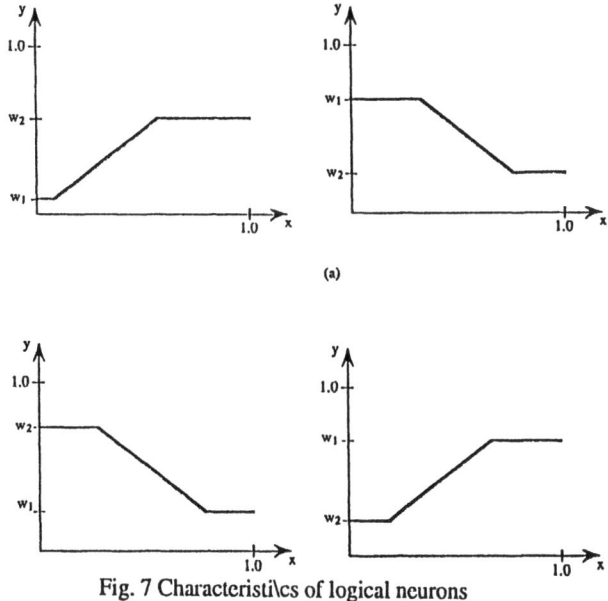

Fig. 7 Characteristi\cs of logical neurons

- The second modification involves an inclusion of a nonlinear transformation following the AND(OR) neuron. Its role is to modify the obtained grades of membership. This functional block does not affect logical properties conveyed by the neuron but performs an additional numerical calibration. The standard two-parametric sigmoid function

$$z = \frac{1}{1 + \exp\left(\frac{-(y-m)^2}{\alpha}\right)}$$

can serve as one of possible instances. The parameter $m \in [0,1]$ and $\alpha > 0$ can easily be adjusted to come up with an appropriate calibration.

The standard s-t composition operator applied to the fuzzy set **x** and the fuzzy relation R has an equivalent representation in terms of "m" OR neurons where each of them possesses "n" inputs. Recall that the basic relational equation can be rewritten as the series of "m" expressions

$$y_j = \overset{n}{\underset{i=1}{S}} (x_i \; t \; r_{ij})$$

j = 1,2,...,m. See also Fig. 8.

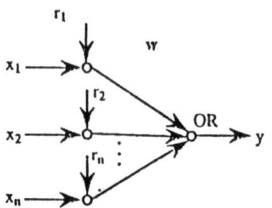

Fig. 8 OR-type logical neuron

The neurons described so far are of an aggregative character. The second type of neurons performs a referential function. Given is a fuzzy reference point, say **r**. Now the neuron is not driven by the input signals x_i themselves but by a collection of degrees of matching achieved for individual coordinates of **x** and **r**. These degrees are afterwards processed by their OR-ing (this could eventually include their weighting by connections w_i). The formalism of this neuron is expressed by the expression:

$$y = \overset{n}{\underset{i=1}{S}} \left[(x_i \equiv r_i) \; t \; w_i \right] \qquad (10)$$

where $x_i \equiv r_i$ is the degree of matching achieved between x_i and r_i. The matching operator, as defined in Section 3, provides an input signal to the OR neuron. We can write it down symbolically as y = MATCH(**x**, **r**).

Thus in fact one can rewrite the neuron described by (10) as a serial structure of the reference block

$$f_i = x_i \equiv w_i$$

followed by the OR neuron

$$y = (f_1 \text{ AND } w_1) \text{ OR } (f_2 \text{ AND } w_2) \text{ OR } ... \text{ OR } (f_n \text{ AND } w_n)$$

The linguistic model can also be formulated in such a way that some other relationships between x_k and **r** are enhanced. We can enumerate two interesting dependences:
- constraint inclusion. The neuron (10) is modified by combining **x** and **r** with the use of the

relationship OR_INCLUDED:

$$y = \text{OR\_INCLUDED}(x, r) \tag{11}$$

In contrast to the previous relationship the level of membership of the output y becomes elevated once x becomes strongly included in the relevant model's constraints. The fuzzy model of this type is driven by the relation of constraint inclusion.
Similarly as before (11) is converted coordinatewise and reads

$$y = [(x_1 \text{ INCLUDED\_IN } r_1) \text{ AND } w_1] \text{ OR } [(x_2 \text{ INCLUDED\_IN } r_2) \text{ AND } w_2] \text{ OR} \\ \ldots \text{ OR } [(x_n \text{ INCLUDED\_IN } r_n) \text{ AND } w_n] \tag{12}$$

The two-argument predicate INCLUDED_IN can be realized by pseudocomplements φ induced by t-norms, see Section 6. If $a < b$ then $a \; \varphi \; b = 1.$, i.e., the constraint "a INCLUDED_IN b" is completely fulfilled.

- The opposite type of dependency induces constraint covering. The grade of membership of resulting y increases as x "covers" r (in other words r is included in x).
This fact is expressed as

$$y = \text{OR\_COVER}(x, r)$$

where the relationship "COVER" is a dual to the predicate INCLUDED_IN, namely

$$y = [(r_1 \text{ INCLUDED\_IN } x_1) \text{ AND } w_1] \text{ OR } [(r_2 \text{ INCLUDED\_IN } x_2) \text{ AND } w_2] \text{ OR } \ldots \\ \text{ OR } [(r_n \text{ INCLUDED\_IN } x_n) \text{ AND } w_n] \tag{13}$$

The structure of the generic neurons (8) - (9) can be enhanced by a bias (threshold) T. Similarily to its role in standard neurons its role here is to carry out thresholding of the output signal. The operation applied here uses the predicate INCLUDED, namely

$$y' = \text{INCLUDED}(T, y)$$

where $T \in [0,1]$ denotes the threshold. If y exceeds T (namely T is included in y) then $y' = 1$ otherwise the value of y' becomes modified by the threshold.

For instance, for the product operation one gets

$$y' = \min(1, \frac{y}{T})$$

and y' is a linear function of y.

For the Godelian φ operator we get

$$y' = \begin{cases} y, & \text{if } T > y \\ 1, & \text{if } T \le y \end{cases}$$

For the above model of the bias we have $y' > y$ for all values of the membership function; for y exceeding T the output value y' is elevated to 1.

7.2 Logical processors as basic processing units

The AND and OR neurons can be put together to form a so-called logic processors (LP). The role of logic processors, which are heterogeneous neural networks, is to realize (approximately) any fuzzy function[22]. Essentially there are two structures of the logic processor:
- The first one consisting of three layers can be viewed as a sum of products (SOM). The input layer consists of "2n" nodes and includes both x_i's as well as their complements ($\bar{x}_i$). The hidden layer includes "p" AND nodes. The output layer has a single OR node.
  Its formal notation looks as follows:
- the hidden layer forms "p" minterms z_j

$$z_j = \underset{i=1}{\overset{2n}{T}} \left(w_{ij} \; s \; x_i' \right) \qquad j = 1,2,\ldots,p \qquad (14)$$

where x' is an extended vector of "2n" inputs including complemented values of all x_i's,

$$x' = [x_1 \; x_2 \; \ldots \; x_n \; \bar{x}_1 \; \bar{x}_2 \; \ldots \; \bar{x}_n]$$

- output layer. The minterms are combined by taking the OR operation on z_j's:

$$y = \underset{j=1}{\overset{p}{S}} \left(v_j \; t \; z_j \right) \qquad (15)$$

- the dual structure of the logical processor computes y by considering a product of minterms and combining the results of the hidden layer by AND-ing them. We will be referring to this structure as a product of maxterms (POM).

Its formal model is described accordingly
- hidden layer

$$z_j = \underset{i=1}{\overset{2n}{S}} \left(w_{ij} \; t \; x_i' \right) \qquad j = 1,2,\ldots,p \qquad (16)$$

- output layer

$$y = \mathop{T}_{j=1}^{p} (v_j \, s \, z_j) \tag{17}$$

The structures of the logical processor are visualized in Fig. 9

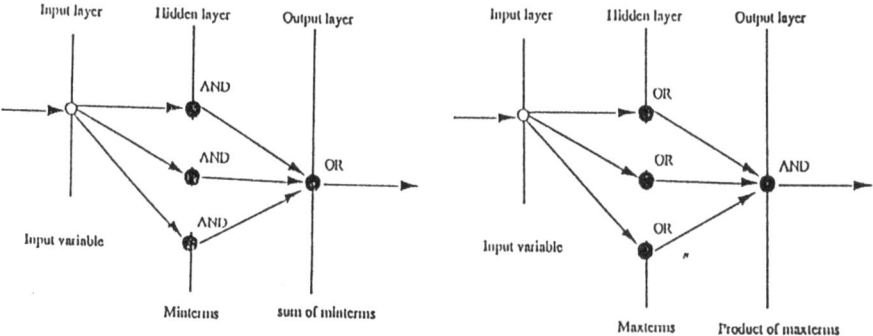

Fig. 9 Structures of the logical processor

The logical processor enables us to approximate logically any continuous function of "n" variables. To handle functions (systems) with "m" outputs one can proceed with "m" separate logic processors operating in parallel.

Another option could be to form a three-layer fully connected network with "2n" input nodes, "p" hidden layers and "m" output nodes. The pertaining formulas are as follows (SOM version)

hidden layer

$$z_j = \mathop{T}_{i=1}^{2n} (w_{ij} \, s \, x_i') \qquad j = 1,2,...,p$$

output layer

$$y_\ell = \mathop{S}_{j=1}^{p} (v_\emptyset \, t \, z_j) \qquad \ell = 1,2,...,m$$

The matching neurons can be also incorporated into the logical processor as a sort of preprocessing unit.

Note now that the size of the input layer of the logical processors is determined by the number of the reference points used in the problem.

7.3 Learning algorithms

The discussed neural networks or even a single neuron require learning. The learning procedures are mainly of a parametric nature and deal with a series of suitable adjustments of weights (connections) of the networks. The learning is carried out on the basis of a learning set of input-output patterns (x_k, y_k), $k = 1,2,...,N$, and is driven by a specified performance index Q. Usually Q is given as the sum of squared errors measuring distances between y_k's and the output of the network while driven by x_k, say $N(x_k)$, where $N(\cdot)$ stands for a general notation of the output of the network obtained for x_k

$$Q = \sum_{k=1}^{N} (y_k - N(x_k))^2$$

The adjustments of connections are driven by a standard Newton-like method. Generally speaking, the abbreviated form of the scheme looks as follows

$$(\text{connections})_{new} = (\text{connections}) - \alpha \frac{\partial Q}{\partial (\text{connections})} \quad (18)$$

while α specifies a rate of learning. This rate implies a suitable speed of learning. Too high values of α could result in oscillations in learning, too small of them could cause a very slow learning.

The general learning formula can be applied to different networks upon specification of all details (such as e.g., triangular norms). Below we summarize the detailed learning scheme for the logical processor described by (14) - (15). We assume that the dimension of the hidden layer is fixed (we will comment later on the selection of this parameter). The triangular norms are also specified in advance, namely we will treat the t-norm as a product while the s-norm will be given as a probabilistic sum[4], namely

$$a \, t \, b = ab \qquad a \, s \, b = a + b - ab$$

The connections to be modified are those between the input and the hidden layer (w_{ij}) and between the hidden layer and the output node (v_j). The general formula (18) is translated into the two following expressions;

$$w_{ij} = w_{ij} - \alpha \frac{\partial Q}{\partial w_{ij}}$$

$i = 1,2,...,2n, \quad j = 1,2,...,p$

---

[4]These triangular norms are also of particular interest in two-valued logic, for Boolean functions developed without Boolean operators cf. [24], Ch. 7.

$$v_j = v_j - \alpha \frac{\partial Q}{\partial v_j}$$
$j = 1,2,...,p$

The computations of the above derivatives are straightforward (below we will concentrate on an on-line type of the learning scheme, where each pair of the training set successively modifies the connections)

$$\frac{\partial Q}{\partial v_j} = \frac{\partial}{\partial v_j}(y_k - N(x_k))^2 = -2(y_k - N(x_k))\frac{\partial N(x_k)}{\partial v_j}$$

And the last derivative is equal to

$$\frac{\partial}{\partial v_j}\left(\underset{\ell=1}{\overset{p}{S}} v_\ell \, t \, z_\ell\right) = \frac{\partial}{\partial v_j}[A + v_j z_j - A v_j z_j] = z_j - A z_j = z_j(1 - A)$$

where $A = \underset{\ell \neq j}{S} v_\ell z_\ell$

For the derivatives of w_{ij} the chaining rule of differentiation applied there yields

$$\frac{\partial}{\partial w_{ij}}(y_k - N(x_k))^2 = -2(y_k - N(x_k))\frac{\partial N(x_k)}{\partial w_{ij}}$$

$$\frac{\partial N}{\partial w_{ij}} \sum_{\ell=1}^{p} \frac{\partial y}{\partial z_\ell}\frac{\partial z_\ell}{\partial w_{ij}} = \frac{\partial y}{\partial z_j}\frac{\partial z_j}{\partial w_{ij}}$$

Finally

$$\frac{\partial y}{\partial z_j} = \frac{\partial}{\partial z_j}\left(\underset{\ell=1}{\overset{p}{S}} v_\ell \, t \, z_\ell\right) = \frac{\partial}{\partial z_j}[A + v_j z_j - A v_j z_j] = v_j(1-A)$$

$$\frac{\partial z_j}{\partial w_{ij}} = \frac{\partial}{\partial w_{ij}}\left[\underset{\ell=1}{\overset{2n}{T}}(w_\emptyset \, s \, x'_\ell)\right] = \frac{\partial}{\partial w_{ij}}[B \bullet (w_{ij} + x'_i - w_{ij} x'_i)] = B(1 - x'_i),$$

$$B = \prod_{\ell \neq i}(w_\emptyset \, s \, x'_\ell)$$

In the logical processor the size of the hidden layer "p" determines its representation capabilities, i.e., uniquely specifies a number of minterms (maxterms) of the logical function the network is capable of handling.

The determination of the size "p" is out of the stream of the above parametric learning. Its choice should be directed by the values of Q and "p" should be increased if necessary.

For the reference neuron and the logic processor driven by the reference preprocessor the reference points can be provided in advance or may be learned as well in the standard parametric manner outlined before.

8. Fuzzy Control In Structures and Algorithms

As it has become clearly visible form the considerations of the previous section fuzzy modelling concerns with modelling at a level of linguistic aggregates. Fuzzy control, on the other hand, is a type of control activity that again is carried out at the same conceptual level. Thus fuzzy models are necessary to work out, analyze and modify control actions by experimenting with them. This type of experimentation will allow to anwer a lot of important questions of "if-then" nature. Note that there is a striking difference between an evident empirical fuzzy control developed for fuzzy controller and the system approach discussed now.

To focus our attention when studying control algorithms we will consider the first order relational model (FR) described as

$$x(k+1) = u(k) \circ x(k) \circ R \tag{19}$$

where u(k) and x(k) denote control and state in the k-th time moment. While x(k+1) stands for the state of the model in the consecutive time instant. Below we will summarize some interesting architectures

1. Fuzzy controller isconstructed as a relational feedback of x(k)

$$u(k) = x(k) \circ T \tag{20}$$

which constitutes a static loop governed by fuzzy relation T.
Insetting (20) into (19) one derives

$$x(k+1) = x(k) \circ T \circ x(k) \circ R$$

Further on, assuming that a certain goal defined as G is provided. The fuzzy feedback relation T has to be determined. More formally we formulate the problem accordingly
- obtain T such that a collection of conditions (equations)

$$G = x(k) \circ T \circ x(k) \circ R \tag{21}$$

is fulfilled for x(k) coming from a finite family of initial state conditions $\mathcal{X}$.

To solve (21) one has to apply an analytical method, see [5] formulate it as an optimization task in which a sum of differences between G and $x(n) \circ T \circ x(n) \circ R$ is minimized.

It is worth noting that (19) is nothing but a standard and a well-known form of the fuzzy controller. The way of its construction is definitely different from that followed in a traditional experimental approach.

Along the same line lies the architecture of the fuzzy controller completed as a series of logic processors of the form

$$u(k) = LP(x(k), w)$$

where the relevant optimization procedure is carried out over $\mathcal{X}$. The fuzzy controller constructed in this way is optimal with respect to all states in $\mathcal{X}$.

2. The second approach involves the fuzzy controller designed by producing its rules. In this case instead of determining the fuzzy relation of the controller one calculates a fuzzy set of control for each x(k) provided. The mode works in an on-line fashion and calls for a solution of the following equation with respect to u(k)

G=u(k) ox(k) oR

3. The fuzzy control is computed in order to satisfy simultaneouly goal G and constraint defined in the space of control C. Inother words v(k) should maximize satisfaction of the two objectives. Since these two objectives are defined in the two different universes of discourse, the fuzzy model is utilized to transform one of them (namely control) into the corresponding state space. An overall scheme in which an optimal control (satisfying G and C) is determined, is visualized in Fig. 10.

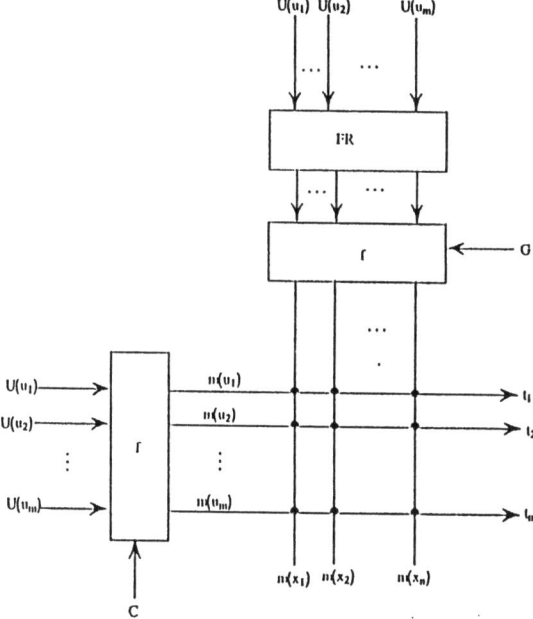

Fig. 10 Determining optimal control for G and C

Let us assume that x(k) is provided. The fuzzy set of control U implies x(k+1). In this scenario x(k+1) = u(k) o x(k) o R which is compared with G. In general the resulting matching operator "f" returns the degree of plausibility (e.g. equality) between these two fuzzy sets. The same holds for u(k) and C. An overall aggregation yields values $t_1, t_2, ...t_m$ where

$$t_i = \sum_{j=1}^{n} m(x_j) \text{ AND } m(n_i)$$

i = 1,2...,m. Then the optimization aims at maximization of the sum $\sum_{i=1}^{m} t_i$ with respect to u(k).

9. Concluding remarks

Fuzzy modelling and identification of fuzzy models occupy a central place in applications of fuzzy sets. We have concentrated on the methodology of fuzzy modelling and thoroughly commented on essential features of applications which called for fuzzy models. The complete identification procedures have been introduced and discussed in length. Some important categories of fuzzy models classified according to their increasing structural content have been proposed. The problem of evaluating fuzzy models and expressing their relevancy has been addressed. Furthermore generic dependences existing between algorithms and recognized in this study are useful in pursuing studies on fuzzy control and fuzzy controllers. This understanding of close relationships and design advantages resulting there will definitely add new dimensions to the research on fuzzy controllers.

10. References

1. B. D'Ambrosio, Qualitative Process Theory Using Linguistic Variables, Springer-Verlag, New York, Berlin, 1989.
2. R.E. Bellman, L.A. Zadeh, "Decision making in a fuzzy environment", Management Science, 17, 1970, 8141 - 81164.
3. J.C. Bezdek, Pattern Recognition with Fuzzy Objective Function Algorithms, Plenum Press, New York, 1981.
4. J.M. Chambers, W.S. Cleveland, B. Kleiner, P.A. Tukey, Graphical Methods for Data Analysis, Wadsworth Int. Belmont, 1983.
5. A. Di Nola, S. Sessa, W. Pedrycz, E. Sanchez, Fuzzy Relational Equations and Their Applications in Knowledge Engineering, Kluwer Academic Press, Dordrecht, 1989.
6. J. De Kleer, J.S. Brown, "A qualitative physics based on confluence", Artificial Intelligence, 24, 7-83, 1984.
7. D. Dubois, H. Prade, Possibility Theory - An Approach to Computerized Processing of Uncertainty, Plenum Press, New York, 1988.
8. P. Eykhoff, System Identification, Parameter and State Estimation, J. Wiley, London, 1974.
9. T. Gilb, S. Finzi, Principles of Software Engineering Management, Addison-Wesley, Reading, MA, 1988.
10. B. Heshmaty, A. Kandel, "Fuzzy linear repression and its application to forecasting in uncertain environment", 15, 1985, 159-191.
11. B.W. Johnson, Design and Analysis of Fault Tolerant Digital Systems, Addison-Wesley, Reading, MA, 1989.
12. R.S. Ledley, Digital Computer and Control Engineering, McGraw Hill, New York, 1962.
13. Menger, K., "Statistical metric spaces", Proc. Nat. Acad. Sci., USA, 28, 535-537, 1942.
14. J.P. Norton, An Introduction to Identification, Academic Press, London, 1986.

15. W. Pedrycz, "On generalized fuzzy relational equations and their applications", J. Math. Anal. Appl. 107, 1985, 520-536.
16. W. Pedrycz, "Processing in relational structures: fuzzy relational equations", Fuzzy Sets and Systems, 40, 1990, 77-106.
17. W. Pedrycz, "Fuzzy sets framework for development of a perception perspective", Fuzzy Sets and Systems, 37, 1990, 123-137.
18. W. Pedrycz, "Direct and inverse problem in comparison fuzzy data", Fuzzy Sets and Systems, 34, 223-236, 1990.
19. W. Pedrycz, "Neurocomputations in relational systems", IEEE Trans. on Pattern Analysis and Machine Intelligence, 13, 1991, 289-296.
20. W. Pedrycz, "Fuzzy modelling: fundamentals, construction and evaluation", Fuzzy Sets and Systems, 41, 1991, 1-15.
21. W. Pedrycz, "Selected issues of frame of knowledge representation realized by means of linquistic labels", Int. Journal of Intelligent Systems, 7, 1992, 155-170.
22. W. Pedrycz, "Fuzzy neural networks with reference neurons as pattern classifiers", IEEE Trans. Neural Networks, 1992 (in press).
23. Ch.J. Puccia, and R. Levins, Qualitative Modeling of Complex Systems, Harvard University Press, Cambridge, MA, London, 1985.
24. S. Rudeanu, Boolean Functions and Equations, North Holland, Amsterdam, 1974.
25. E. Sanchez, "Resolution of composite fuzzy relation equations", Information and Control, 34, 38-48, 1976.
26. D.A. Savic, W. Pedrycz, "Evaluation of fuzzy linear regression models", Fuzzy Sets and Systems, 39, 1991, 51-63.
27. W.G. Schneeweiss, Boolean Functions with Engineering Applications, Springer Verlag, Berlin, 1989.
28. M. Sugeno (ed.), Industrial Applications of Fuzzy Control, North Holland, Amsterdam, 1985.
29. T. Takagi, M. Sugeno, "Fuzzy identification of systems and the applications to modelling and control", IEEE Trans. on Syst., Man and Cybern, 15, 1985, 116-132.
30. H. Tanaka, S. Uejima, K. Asai, "Linear repression analysis with fuzzy model", IEEE Trans. Syst. Man and Cybern, 6, 1982, 903-907.
31. H. Tanaka, "Fuzzy data analysis by possibilistic linear models, Fuzzy Sets and Systems, 24, 1987, 363-375.
32. I.B. Turksen, "Measurement of membership and their acquisition", Fuzzy Sets and Systems, 40, 1991, 5-38.
33. R.R. Yager, On measuring specificity, Tech. Rep. Iona College, New Rochelle, NY, 1990.
34. L.A. Zadeh, "Fuzzy sets and systems", Proc. Symp. Syst. Theory Polytech. Inst. Brooklyn, 1965, 29-37.
35. L.A. Zadeh, "Toward a theory of fuzzy systems", In: Aspects of Network and System Theory (R.E. Kalman, N.De Claris, eds), 1971, 469-490.
36. Zadeh, L.A., "Fuzzy sets", Information and Control, 8, 338-353, 1965.

37. L.A. Zadeh, "A rationale for fuzzy control", Trans. ASME, Ser. G, 94, 1974, 3-4.
38. L.A. Zadeh, "PRUF-a meaning respresentation language for natural language", Int. J. Man - Machine Studies, 10, 1978, 395-346.
39. L.A. Zadeh, "Fuzzy sets as a basis for a theory of possibility", Fuzzy Sets and Systems, 1, 1978, 3-28.
40. Zadeh, L.A., "Outline of a new approach to analysis of complex systems and decision processes", IEEE Trans. Syst., Man and Cybern., 1, 28-44, 1973.
41. Zadeh, L.A., "Fuzzy sets and information granularity", in: M.M. Gupta, R.K. Ragade, R.R. Yager, eds., Advances in Fuzzy Set Theory and Applications, North Holland, Amsterdam, 3-18, 1979.

The Representation and Use of Uncertainty and Metaknowledge in MILORD, a System for Diagnostic Reasoning

R. López de Mántaras

Artificial Intelligence Research Institute
Centre d'Estudis Avançats de Blanes, Camí de Santa Bàrbara. 17300 Blanes, Girona, Spain.
e-mail: mantaras@ceab.es

Abstract

In this paper we present the management of uncertainty proposed in the COLAPSES language. Uncertainty is mainly a control parameter of the execution of a knowledge base. This language allows the possibility of defining local logics inside a hierarchic structure of modules containing rules. These local logics interact between them in a user defined fashion that gives to the uncertainty treatment a flexibility not found in other systems. The relation between control (meta-logic) and deduction (logic) is based on a reflexive architecture that uses introspection mechanisms to take decisions from the uncertainty in the facts. These features allow to mimic more closely human problem solving behaviour and particularly when uncertainty is present. We conjecture that this type of architecture may have implications for ITS because it allows to provide explanations involving control knowledge decisions.

1. Introduction

Most AI research on reasoning under uncertainty is concerned with global normative methods to propagate and combine certainty values and there is a controversial debate about which methods are most appropiate and why. Disagreement between the proponents of the different methods (Bayesians, Dempster-Shafer, Fuzzy logicians) is about the meaning of uncertainty and having a formalism that produces rational conclusions with no claim to mimic human uncertainty management methods. In restricted domains where the uncertainty involved permits, for example, a direct interpretation in the probability theory sense and where the main decision of the expert system involves the computation of such uncertainty if appropiate data are available, then probability calculus would be the best independently of how humans solve problems in the same situation. However, the most interesting aspect of the expert systems research is to gain some insights of human problem solving strategies by trying to emulate them in programs. Although human problem solvers are almost always uncertain about the possible solution, they very often achieve their goals despite uncertainty by using methods to manage uncertainty that are particularized to the type of problem solving that they are performing at a given time. In fact, we believe that managing uncertainty consists in selecting actions that simultaneously achieve solutions and reduce their uncertainty. Since actions are selected not only for their domain effects but also for their effect on uncertainty, problem solving under uncertainty is more constrained that problem solving under total certainty as was also noticed by Cohen [2-3].

This view leads to consider uncertainty mainly as a control feature because it helps to constrain the focus of attention (i.e. which part of the problem to work next) and action selection (i.e. how to work on it). The problem solving strategies are implemented in the

---

[1]Research partialy supported by CICYT project SPES nº 880j382

problem solver's control part. The knowledge engineers translate into control strategies the human problem solving strategies.

Problem solving strategies are a fundamental component of expertise and have to be acquired by first implementing a set of task-level primitives with which experts can describe their strategies. Uncertainty is, in our opinion, a task-level primitive that is used at the implementation level to descriminate alternative control decisions. Furthermore, when large expert systems emulate human problem solving strategies, the organization of their complex knowledge bases makes the propagation and combination of uncertainty a local, context dependent, process. In our opinion, such large domain expert systems draw their problem solving capabilities more from the power of their organizational and problem solving structures than from the particular uncertainty management formalism they use (different formalisms can be adjusted to give similar answers).

This paper discusses these ideas in the framework of COLAPSES a modular language based upon the MILORD language [4]. This expert systems building tool uses uncertainty as a control feature and performs local combination and propagation of uncertainty. A medical diagnosis application is used as example whose potential tutoring capabilities, to perform case studies by advanced medical students, are enhaced by the capability to provide justifications of the problem solving strategies selected at the control level.

2. Modularity and locality

A knowledge base is a large set of knowledge units that cover a domain of expertise and provide solutions to problems in that domain of expertise.

When faced with a particular case, human experts use only a subset of their knowledge for two reasons: adecuacy of the general knowledge -the theory- to the particular problem and availability and cost of data. For exemple, the suspicion of a bacterian disease will rule out all knowledge referring to virical diseases; and also a patient in coma will make useless all the knowledge units that need patient's answers.

The adequacy of general knowledge to a particular problem is done at a certain level of granularity, for instance, the expert uses all the knowledge related to the diagnosis of a colon neoplasy or the knowledge related to the radiological analysis of a chest x-ray.

In particular the structuration of KB's is made in MILORD taking into account this granularity in the use of knowledge.

Each structural unit or theory (module from now on) will define an indivisible set of knowledge units (for example rules and predicates). The control will be responsible for the combination of the modules. The combinations will represent the particularization of general knowledge to the problem that is being solved. The control will determine which combinations are acceptable. For example, a module that determines the dosis of penicillin that has to be given to a patient must not be present in any acceptable combination for a patient allergic to penicillin.

The modularization of KB's leads to the concept of locality in the modules of a KB. It is possible to define the contents of a module independently of the definition of the rest of the modules. This possiblility, methodologicaly desirable, allows the use of different local logics and reasoning mechanisms adapted to the subtasks that the system is performing.

2.1 Modularity over MILORD: The COLAPSES language

The basic units of KB's written in our language, COLAPSES, are the modules. These may be hierarchically organized, and consist of an excapsulated set of import, export, rule, meta-rule and submodule declarations. The declaration of submodules in a module is what structures the hierarchy. The declaration of submodules does not differ from the declaration of modules. We shall briefly outline which is the meaning of the primitive components of a module. A complete definition of the language and its semantics can be found in [5].

Import: determines the non-deducible facts needed in the module to apply the rules. These facts are to be obtained from the user at run time.

Export: defines which facts deduced or imported inside a module are visible from the rest of the modules that include the modules as a submodule. The import, submodule declarations and export components define the interface of the module.

Rule: defines the deductive units that relate the import and the export components within a module.

Metarule: defines the meta-logical component of the module. Thus, the meta-rules of a module will control the execution of the rules in the module and the execution of the submodules in the hierarchy underneath the module.

The syntax of a module definition is as follows:

Module *modid* = *modexpr*

where *modid* stands for an identifier of the module and *modexpr* for the body of the definition made out of the components specified above. Let us look at an example of module definition.

Module Gram_esputum =
  begin
      import Class, Morphology
      export Morphology, Esputum_ok
      deductive knowledge:
          Rules
          R001 **if** Class > 4 **then** Esputum_ok **is** sure
          ...
      end deductive
  end

There is also the possibility of defining generic modules that represent functional abstractions of several non generic modules.

2.2. Local Logics

It is clear that the experts use different approaches to the management of uncertainty depending on the task they are performing. Usually expert system building tools provide a fixed way of dealing with uncertainty proposing a unique and global method for representing and combining evidence. In the COLAPSES language it is possible to define different deduction procedures for each one of the modules. If from a methodological point of view a task is associated with a module then, a different logic can be used depending on the task. The definition of local logics is made by the next primitive in the COLAPSES language:

Inference systems
    Truth values = *list of linguistic terms*
    Renaming = *morphisms between linguistic terms*
    Connectives:

Inference patterns. This field defines the inference rules to be used along the deductive process. To date the only accepted pattern in COLAPSES in the modus ponens for which a propagating function can be defined.

3. Meta-reasoning by introspection using uncertainty

Having considered uncertainty as a logical component of the COLAPSES language, i.e. the semantics of formulae, the control of reasoning under the uncertainty must be considered as a meta-logic component. Thus the meta-inference over the uncertainty will determine which will the inference control be at the logic level. This meta-inference acts upon the logic component using mechanisms of introspection, this is, the same language represents the uncertainty of the propositions and provides mechanisms both to look at this uncertainty and to determine the control to be followed.

This meta-control is defined as a component of the modules, allowing a local meta-logic defintion. This control component acts over the deductive knowledge and over the submodule hierarchy. It determines which rules and submodules are useful for the current case. The mechanism of interaction between both components is a reflection mechanism: the deductive component reflects on the control component to know which will be the next strategic step, which submodule to execute next, or which rule to use next.

It is not a full reflection mechanism because we allow the meta-logic to see only the valuation of atomic formulae (facts) and the valuation of strategies (sets of modules that combined can lead the system to the solution of the problem), rules and meta-rules cannot be consulted by the meta-logic.

This general mechanism is used to drive the inference process in different directions; we are going to discuss some of them.

3.1. Evidence increasing

The current uncertainty of facts can be used to control the deduction steps in order to increase the evidence of a given hypothesis. So, for example, if we have an alcoholic patient showing a cavitation in the chest x-ray and there is low evidence for tuberculosis, then the Ziehl-Nielssen test to determine more clearly whether he has a tuberculosis should not be done. But if he also presents a risk factor for AIDS then we shall increase our evidence for tuberculosis and the test will be suggested. This is expressed as follows:

If tuberculosis > moderately_possible
then conclude Test Ziehl-Nielssen

If risk_factor_for_AIDS **then conclude** tuberculosis **is** possible

If Alcoholic **and** Cavitation
then tuberculosis **is** almost_impossible

It should be noticed that the first rule is a rule of the meta-logic component of the language whilst the others are rules at the logic level.

3.2. Strategy focusing

The uncertainty of facts can determine the set of hypothesis to be followed in the sequel. Example:

> **If** the pneumonia is bacterial with certainty < quite_possible **and**
> the pneumonia is atipical with certainty > possible
> **Then focus on**
> Mycoplasm, Virus, Clamidia, Tuberculosis, Nocardia, Criptococcus, Pneumocistis-Carinii
> **with certainty** quite possible

This example means that the modules to be used in order to find a solution to the current case are those indicated in the conclusion of the meta-rule and should be considered in the order specified there.

Strategies have a certainty degree attached to them. This is useful to differentiate the strategies generated by very specific data from those generated by general data. As an example consider the case of a patient with AIDS (a kind of immunodepression). If we know that the patient suffers from AIDS, a more specific strategy (and also more certain) can be generated. But if we just know that the patient has an immunodepression a less certain general strategy would be generated. Since we may have several candidate strategies simultaneously, combining different strategies is a matter of great importance in the control of the system. This is also achieved by looking at the uncertainty of the strategies, as shows the next example:

> **If** Strategy (X) **and** Strategy (Y) **and** Certainty (X) > Certainty (Y) **and**
> Goals (X) $\cap$ Goals (Y) $\neq \emptyset$
> **Then** Ockham (X,Y)

where Ockham (X,Y) is a combination of the strategies that gives priority to those modules found in the intersection of both strategies: (Goals (X) $\cap$ Goals (Y))

3.3. Knowledge adequation

As indicated at the begining of the paper a KB is a set knowledge units that have to be adapted to the current case. For example alcoholism is a useful concept when determining a bacterial pneumonia, but it is useless for non-bacterial diseases. Then, a possible use of the uncertainty of the fact bacterianicity is to decide about the use of a given concept in the whole KB, i.e. to adequate the general knowledge to the particular problem. Example:

> **If** no bacterian disease
> **then** do not use alcoholism in finding the solution

3.4. Solution acceptance

The degree of uncertainty of a fact can also be used to stop the execution of the system. For example

> **If** Pneumocistis-carinii **and** tuberculosis < possible **and** Criptococcus < possible
> **Then stop**

The control tasks we have discussed use uncertainty as a control parameter and are tasks of the meta-logic level. They are represented as a local meta-logic component of each module in what is called the control knowledge component of a module.

4. Metacontrol and locality

The structured definition of KB's helps not only in the definition of safe and maintainable KB's but also gives some new features that where impossible to achieve in the previous generation of systems. Among them the most important is the possibility of defining a local meta-logical component for each one of the modules.

The definition of strategies (ordered set of elementary steps to solve a problem) in the MILORD system was made globally. Only one strategy could be active at any moment. Presently, as many strategies as nodes in the modules graph structure can be active. This flexibility is linked with the fact that each module can have a different treatment of uncertainty. So, uncertainty plays a different role as a control feature depending on the association between module and logic.

Furthermore, given the fact that the system consists of a hierarchy of submodules the meta-logical components act ones upon the others in a pyramidal fashion. This allows us to have as many meta-logic levels as necessary in an application. Further research will be pursued along this line. A richer representation of the logic components in the meta-logic will also be investigated and sound semantics from the logic point of view will be defined. Finally, the potential for tutoring advanced medical students due to the higher explanation capabilities involving control knowledge will be evaluated.

5. Conclusions

The interesting aspect of building expert systems is to learn something about human problem solving strategies by trying to reproduce them in programs. Human problem solver's are uncertain in many situations and do not use a simple normative method to handle uncertainty. Instead they take advantage of a good organization in the problem solving task to obtain good solutions using qualitative approximations. This suggests to consider uncertainty as playing an important role at the control level by guiding the problem solving strategies. In order to illustrate these points, we have shortly described a modular architecture and language that extensively exploits uncertainty as a control feature and uses local context dependent combination and propagation uncertainty operators

6. Bibliography

1. Agustí J., Sierra C., Sannella D. (1989): "Adding generic modules to flat rule-based languages: a low cost approach", in *Methodologies for Intelligent Systems 4*, (Z. Ras, ed.), Elsevier Science Pub., pp. 43-51.
2. Cohen P.R., Day D., De lisio J., Greenberg M., Kjeldsen R., Suthers D., Berman P. (1987): "Management of Uncertainty in Medicine", *International Journal of Approximate Reasoning* 1: 103-116.
3. Cohen P.R. (1987): "The Control of Reasoning Under Uncertainty: A Discussion of Some Programs", COINS Technical Report 87-81, University of Massachusetts at Amherst.

4. Godo L., López de Mántaras R., Sierra C., Verdaguer A., (1987): "MILORD, the architecture and management of linguistically expressed uncertainty", *International Journal of Intelligent Systems*, Vol. 4, nº 4, pp. 471-501.
5. Sierra C., Agustí J. (1990): "COLAPSES: Syntax and Semantics", CEAB Research Report 90/8.

Approximate Reasoning And Control With Sparse And / Or Inconsistent Fuzzy Rule Bases

László T. Kóczy*

Department of Computer Science, POSTECH
P.O.Box 125, Pohang 790-600, Korea

1 Introduction

Fuzzy theory offers a very convenient way to formalize approximate reasoning. There are many problems where the system to describe is too complex to fit any reasonable and tractable conventional exact model. And still, it is possible to manage such systems to a certain degree, with an acceptable suboptimality, like it is done when a medical doctor diagnoses a patient without knowing too much about the incredibly complex processes in the human body and also, when a well experienced operator sets the input values for a highly nonlinear industrial system with eventually several hundred parameters just by feeling.

Such experts can usually formulate the essence of their knowledge by listing a large amount of rules. These rules can always be formalized in the way "If the observation is A, then the decision or action is B". The rules contain linguistic expressions with vague notions and the rules often do not cover the full range of possible observations. For the doctor, every patient is different, he can virtually never observe exactly the same constellation of symptoms – and still, he can conclude in most cases with a high accuracy. The reason is that he uses some kind of instinctive generalization technique, he compares the situation with many similar cases and he finds some rather analogous ones, where the conclusion is already known. A similar behaviour is typical

---
*On leave from the Dept. of Telecommunication and Telematics, Technical University of Budapest, Sztoczek 2, Budapest H-1111, Hungary

also for the well experienced operator at the industrial plant. He recognizes the similarity of the observed situation with some previously known situations and, although the configuration of parameters is very likely never exactly the same, he is able to extract the dominating features of the situation and so he can find similar cases in his knowledge base where succesful decisions had been taken, and in this way he finds some compromise which leads to another good action.

Real life produces sometimes even more complicated cases of decision. When several doctors consult over a particularly difficult combination of symptoms they unite their knowledge. However their experience is consonant, some conflict is always expected, because of the unexactness of linguistically formulated rules representing their experience and because of the nature of the object of their study. This kind of conflict is usually not very deep, contradiction in the experience does not exceed the limits of a "band", and by averaging the knowledge – sometimes by filtering out extremal pieces of information conflicting with the rest of the rules – some good decision can be taken. The same is true for many engineering problems, where the complexity of the system exceeds the limits of the exactly describable.

This paper deals with the special aspects of fuzzy reasoning and control where the available knowledge (in the form of "If...then" style rules) does not cover the input space densely enough, and/or it contains some internal contradictions like when attaching more than one different consequents to the same antecedent. Finally, it touches the subject of structured hierarchical control which is especially interesting in the case when systems with a very high number of (input) variables or otherwise very complex structure are to control.

2 Reasoning by the Compositional Rule of Inference

Fuzzy rule based reasoning and control have come to the focus of interest of many computer, control and systems scientists because of the large amount of recent industrial applications, especially in Japan. The basic idea was proposed by Zadeh [1], when he suggested the description of highly complex systems by "If...then..." rules and proposed the formal interpretation of such a rule as a vague formulation of the function connection between input and output variables, rather than a logical implication. Soon, the practical implementation appeared, with some simplified computational techniques

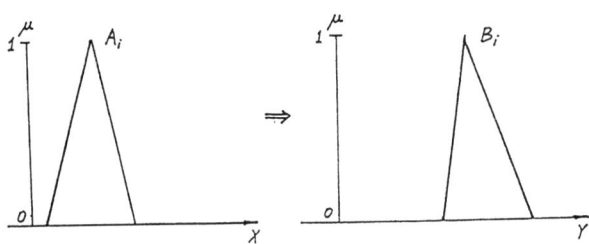

Figure 1: A simple 1+1 dimensional rule "If X is A_i then Y is B_i expressed by fuzzy sets over X and Y

introduced by Mamdani *et al.* [2]. Reason for modifying the algorithm was the high (exponential) computational complexity of the original method. It is not easy to understand why it took more than one decade – despite very promising initial results – until really masswise applications appeared, virtually only around and after the 2nd IFSA World Congress held in 1987 at Tokyo, Japan.

The Compositional Rule of Inference uses a set of rules in the form

"If X is A_i then Y is B_i"

where in most cases both X and Y are compound variables, i.e. $X = X_1 \times X_2 \times ... \times X_m$ and similarly $Y = Y_1 \times Y_2 \times ... \times Y_n$. Then, also the antecedent A_i and the consequent B_i are composed in multiple dimensions. A_i and B_i are linguistic terms expressing the approximate value of their respective variables, as e.g. if X is velocity, A_i might be "very low". Linguistic terms can be represented as fuzzy sets of the universe of all exact values corresponding to the given variable, in our example, 0 to 250 km/h if the velocity is understood in the context of a normal car. So "very low" means something like 0 to 10 km/h, e.g. Such rules can be represented by fuzzy sets over X_i and Y_j, like on Figure 1, where both input and output are one dimensional.

This set of rules forms a rule base $R = \{R_1, ..., R_r\}$. An observation or fact has always the form

"X is A^*"

and it can be interpreted as a fuzzy (or, sometimes, crisp) value of X. The rule base is combined with this observation in such a way that the most likely conclusion is calculated by analogy of the rules containing antecedents more or less similar to the observation. Formally, this is done by intersecting the cylindric extension of the observation and the union of the relations

representing the rules. The resulting membership function is projected to Y and so a conclusion
"Y is B^*"
is obtained:

$$B^*(y) = sup_x\{min\{A^*(x), max_i\{min\{A_i(x), B_i(y)\}\}\}\} \qquad (1)$$

Because of the exponential complexity of such a calculation (X is usually multidimensional!), this method cannot be used directly except for very low numbers of (input) variables. The actual technique of inference is usually the calculation of some kind of degree of overlapping ("similarity") between A^* and A_i. Those rules where this overlapping is greater than zero (i.e., $supp(A_i) \cap supp(A^*) \neq \emptyset$) are fired with A^*, and their conclusion is weighted by the degree of overlapping, usually expressed as the supremum (height) of the intersection of the observation and antecedent. These weighted or truncated consequents are unioned separately in every component of Y and so they result in the components of B^*. The above technique can be expressed formally as

$$w_i = min_j\{sup_{x_j}\{min\{A^*(x_j), A_i(x_j)\}\}\}$$

$$B^*(y_k) = max_i\{min\{w_i, B_i(y_k)\}\}$$

or

$$B^*(y_k) = max_i\{w_i B_i(y_k)\}$$

and

$$B^*(y) = \prod_{i=1}^{r} B^*(y_k) \qquad (2)$$

Clearly, (1) and either version of (2) do not generate the same result!

3 Dense and sparse rule bases

The above methods of inference are built up on some fuzzy set or single membership degree expressing the overlapping of observation and rules in (1), then the intersection of the observation and the union of rules is projected to Y. In (2), it is the minimal height of intersection selected from the components of X that limits ("truncates") or compresses the consequent

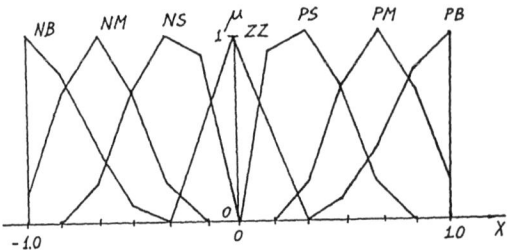

Figure 2: Antecedents "Negative Big", "Negative Medium", Negative Small", "Zero", "Positive Small", "Positive Medium" and "Positive Big" in the control application of Mamdani et al.

functions. In the classical control applications of fuzzy approximate reasoning, the rule bases are constructed always in such a form that the union of supports of antecedents is equal to the input space:

$$\bigcup_{i=1}^{r} supp(A_i) = X$$

even, for a reasonably high $\alpha \in (1, 0]$, the α-cuts still cover X:

$$\bigcup_{i=1}^{r} A_{i\alpha} = X$$

Figure 2 presents the set of antecedents used in the very first application in [2] used to describe the linguistic degrees applied as antecedents (in this example: change in pressure error, we normalized the input range for [-1,+1]). It is clear that for $\alpha \simeq 0.65$ still there is no such $x \in X$ that it is not in the α-cut of at least one A_i.

In many recent approaches, the membership functions of the antecedents have a very simple triangular shape. A typical set of linguistic terms used in many fuzzy control systems is presented on Figure 3, the set of antecedents is taken from [3].

Such a situation can be described as the case when the antecedents $A_1...A_r$ form a *fuzzy partition*, or more exactly, an α-*partition* of X. Figure 4 depicts a dense rule system in the cross product space $X \times Y$, by the supports of the rules. As mentioned above, a similar picture is generated if the α-cuts are considered for a reasonably high α. The support of R_i is the cross

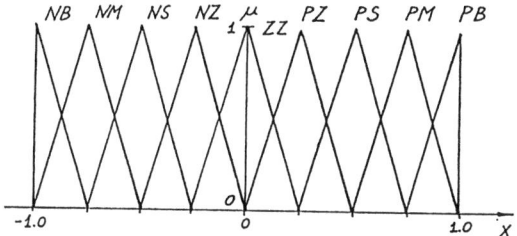

Figure 3: Typical fuzzy partition of X by triangular antecedents

product of $supp(A_i)$ and $supp(B_i)$. Note that while the rules form a partition of X (supposed to be bounded from below and above), it is not necessarily true that they form a partition of Y, as well. Depending on the range of Y in Figure 4, the union of supports of rules might be a proper subset of Y only.

In contrary to the dense case, when the parametrized tuning technique in [3] is applied, the distribution of A_i changes, as e.g. on Figure 5. Some of the terms representing antecedents ("Negative Small", "Negative Zero", "Zero", "Positive Zero" and "Positive Small") are "compressed" around 0 (in the interval $[-s, +s]$, where s is the tuning parameter), while the rest remains at the original place. In such a way, some areas of X (on the figure, the intervals $[sup\{supp(NM)\}, inf\{supp(NS)\}]$ and $[sup\{supp(PS)\}, inf\{supp(PM)\}]$ remain completely uncovered).

What happens if the observation lies in those areas? Figure 6 presents such a situation in the support view: the support of the cylindric extension of A^* intersects with no rule antecedent.

In sparse rule bases both equations (1) and (2) result equally $B^*(y) = 0$. However, in many practical situations, it is not possible to establish rules densely over X, lack of information or even computational complexity might motivate the generation of sparse antecedents (cf. [4]). Our purpose is to find some way to cope with such problematic rule bases, where still inference by careful evaluation of *neighbouring* antecedents might be possible!

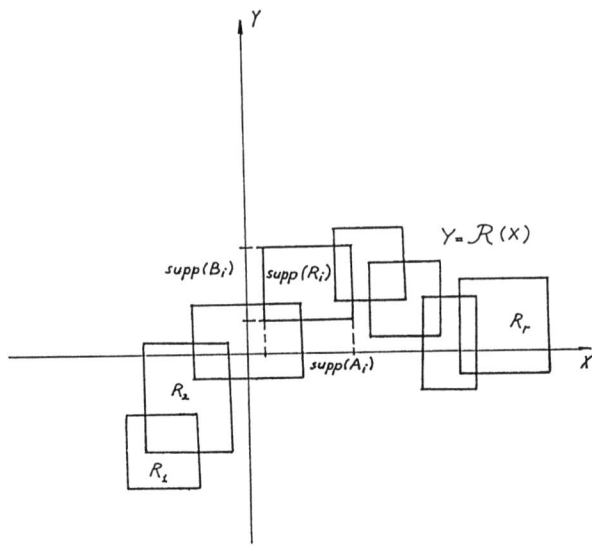

Figure 4: Support view of a dense rule system covering the full range of the input space

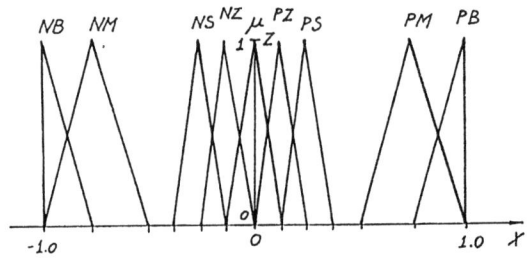

Figure 5: Linguistic antecedent set after parametrized tuning of Burkhardt et al.: generating a sparse rule system

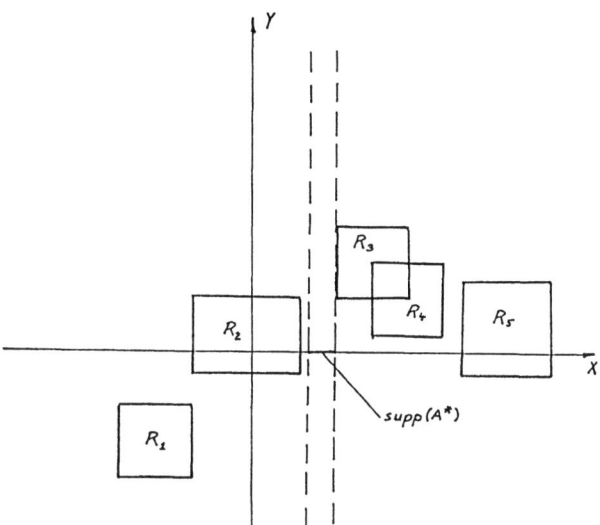

Figure 6: Sparse rule system in the support view: the cylindric extension of the observation does not intersect with any antecedent

4 Gradual rules, gradual variables and analogical reasoning by similarity

In order to deal formally with sparse rule bases, an extended interpretation of CRI must be defined. There are three approaches in the literature which contain various elements of dealing with this kind of problem. The semantical interpretation of "If...then..." rules as *gradual rules* in [5] focuses on the fact that antecedent and consequent express properties of the input and output variables, which are present to a certain extent, and any observation can be evaluated from the point of view of its being more or less like a given antecedent. According to this approach, the rule can be interpreted as
"*The more X is A_i the more Y is B_i*".
It must be seen clearly that such an interpretation is possible only if the *variables themselves are gradual*, i.e. it is possible to speak about "more or less" in the sense of X and Y. This condition is always satisfied in industrial control applications as the variables occuring there — pressure, temperature, position, change of these, etc. — are measurable and have a natural ordering which generates a range from the minimal to the maximal degree of

the property. Even in a context of an expert system where physically non measurable variables occur like "beautiful", "ripe" or "cloudy", it is possible to establish an ordering, sometimes by merely subjective evaluation, sometimes by the help of some objective feature (as e.g. the percentage and distribution of clouds in the sky).

The second approach is the *analogical reasoning by similarity* in [6]. The essential idea is that some kind of *distance* between two overlapping fuzzy sets is defined and a similarity index (decreasing with increasing distance) between the observation and each of the antecedents is calculated to express the weight of a certain rule in the calculation of the conclusion. There are several notions of the distance of fuzzy sets in the literature, some of them are suitable to be used as the base for similarity. However, the examples given here are referring to partially overlapping pairs and define similarity as equal to 0 if the supports are disjoint. The analogical reasoning technique can be formulated in a similar verbal way as the gradual rule interpretation as follows:

"The more similar is A^* to A_i the more similar should be B^* to B_i".

5 Interpolation and extrapolation in rule bases

The third approach is the family of interpolation/extrapolation techniques based on the Resolution Principle (cf. [7, 8, 9]). In this approach, it is essential that the variables are not only measurable (at least in the subjective sense) but their range can be mapped into a normalized scale where *distance* exists. With the example of car speed, the range $[0km/h, 250km/h]$ is mapped into $[0,1]$ and so the normalized distance of e.g. $v_1 = 60km/h$ and $v_2 = 100km/h$ is $(100km/h - 60km/h)/250km/h = 0.16$. If ripening tomatoes are evaluated according to their colours, the distance of a red and a yellow tomato is roughly 0.5, as "yellow" is rather in the middle of the colour scale in this context while "red" might represent the upper limit of the scale.

It is also necessary that the linguistic terms used in the rules be nomal (i.e. $height(A_i) = height(B_i) = 1$) and convex (for every $x_1, x_2 \in X$ and $\lambda \in (0,1)$, $\mu_{A_i}(\lambda x_1 + (1-\lambda)x_2) \geq max\{\mu_{A_i}(x_1), \mu_{A_i}(x_2)\}$). This condition is however naturally satisfied in the practical problems.

Distance can be introduced as a fuzzy set of normalized distances rather than a scalar value. By the Resolution Principle, the distance on level α represents a value in the universe of distances which has a membership degree

α in the fuzzy distance set. There are several ways of defining the α level distance, as e.g. the distance of the nearest two points of the α-cuts, or the medium distance of α-cuts (i.e. the distance of the two centers of the cuts). In the case of normal and convex A_i and B_i, the α-cuts of rules are hyperintervals (oblongs in the case of one dimensional X and Y), and hyperintervals can be always unambiguously described by their minimal and maximal points $((inf\{supp(A_i(x_1))\}, inf\{supp(A_i(x_2))\}, ..., inf\{supp(B_i(y_n))\})$, and similarly, the maximal point with sup), so it is convenient to describe the distance of two fuzzy sets by two fuzzy sets of distances: the respective distance of the infimum ("lower") points and supremum ("upper") points. If the distance is normalized in every X_i and Y_i then the hyperspatial Euclidean distance of two fuzzy sets R_1 and R_2 is always in $[0, \sqrt{m+n}]$ (where m is the number of input, n the number of output variables, i.e. $m+n$ is the number of dimensions of the $X \times Y$ hyperspace).

On the base of this fuzzy distance, it is possible to reinterpret the idea of gradual and analogical reasoning by the more concrete notion of "nearness" or "closeness":

'The nearer is A^* to A_i, the nearer should be B^* to B_i'.

"Nearness" is decomposed into "α-nearness", i.e. the above semantic interpretation contains theoretically infinite many rules, according to the cardinality of the level set Λ (the set containing all α's occuring in the range of the membership functions of the antecedents and consequents). However, in most applications, the membership functions are piecewise linear (triangular, trapezoidal, etc), as it can be seen on Figures 2 and 3, so it is sufficient to restrict Λ to the "corners" in μ, e.g. in the case of trapezoidal (and triangular) functions $\Lambda = \{0, 1\}$. In this way, the original rules are decomposed into $|\Lambda|$ "α-rules":

'The nearer is A_α^* to $A_{i\alpha}$, the nearer should be B_α^* to $B_{i\alpha}$'

and the membership function of B^* is obtained by the Resolution Principle as

$$B^* = \bigcup_{\alpha \in \Lambda} \alpha B_\alpha^*$$

where B_α^* is calculated by the respective nearness of the α-cuts.

On the base of these ideas it is possible to introduce various methods of approximating the α-cuts of the conclusion. The simplest idea is the *interpolation of two flanking rules*, i.e. if the observation does not overlap with the rules, the two nearest antecedents, one on each side, are selected, and the conclusion is calculated by

$$D(A_\alpha^*, A_{1,\alpha}) : D(A_\alpha^*, A_{2,\alpha}) = D(B_\alpha^*, B_{1,\alpha}) : D(B_\alpha^*, B_{2,\alpha}) \qquad (3)$$

Condition of applicability of this equation is that the two antecedents are separated by the observation, i.e. for every α, the lower and upper points of the α-cuts of the observation lie between the same points of the antecedents, further on that the α-cuts of the consequents are ordered in the same sense.

As we mentioned above, the α-cuts of rules are hyperintervals. As any hyperinterval can be described unambiguously by two characteristic points (the minimal and maximal or "upper" and "lower" points), these two points are selected for the linear interpolation, too. As a matter of course, the interpolation must be done independently for every α-cut, however, if Λ has a finite number of elements only (as e.g. 2 in the case of trapezoidal or triangular rules), altogether $2|\Lambda|$ interpolations are sufficient for reconstructing the entire membership function of the fuzzy conclusion – based on the interpretation of the rules as representative points of the fuzzy function $y = \mathcal{R}(x)$, a fuzzy value in Y, i.e. a membership function directly in the form $B^*(y)$. On Figure 7, this idea is depicted. The two dimensional intervals R_i can be interpreted as supports or arbitrary α-cuts of the rules. On the figure, the support case is indicated: the support of A^* lies between the supports of rules R_2 and R_3, so the lower and upper points of these rules are interpolated, the support of the conclusion is found as the projection of the interval $A^* \times B^*$ to the Y axis. On the figure, both points are interpolated, a clearer overview is given if for every different $\alpha \in \Lambda$ and lower/upper point another graph is drawn.

The exact formulas for the interpolation can be determined by the solution of equation (3):

$$inf_\prec\{B_\alpha^*\} = \frac{\frac{1}{\bar{d}_{\alpha L}(A_{1,\alpha}, A_\alpha^*)} inf_\prec\{B_{1,\alpha}\} + \frac{1}{\bar{d}_{\alpha L}(A_\alpha^*, A_{2,\alpha})} inf_\prec\{B_{2,\alpha}\}}{\frac{1}{\bar{d}_{\alpha L}(A_{1,\alpha}, A_\alpha^*)} + \frac{1}{\bar{d}_{\alpha L}(A_\alpha^*, A_{2,\alpha})}}$$

$$sup_\prec\{B_\alpha^*\} = \frac{\frac{1}{\bar{d}_{\alpha U}(A_{1,\alpha}, A_\alpha^*)} sup_\prec\{B_{1,\alpha}\} + \frac{1}{\bar{d}_{\alpha U}(A_\alpha^*, A_{2,\alpha})} sup_\prec\{B_{2,\alpha}\}}{\frac{1}{\bar{d}_{\alpha U}(A_{1,\alpha}, A_\alpha^*)} + \frac{1}{\bar{d}_{\alpha U}(A_\alpha^*, A_{2,\alpha})}} \qquad (4)$$

Here $\prec$ stands for the partial orderings in X and Y, and $inf_\prec$ for the infimum in the sense of $\prec$, etc. $\bar{d}$ denotes the fuzzy distance between two fuzzy sets.

The solutions expressed in this way give some ideas concerning a more general approach. If more than two rules are considered, all of them can be

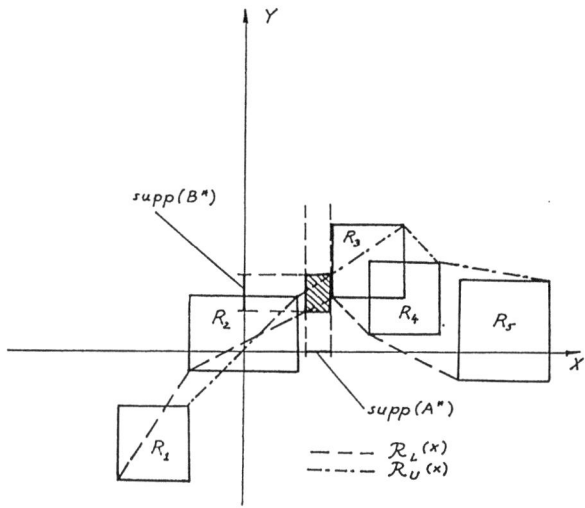

Figure 7: Interpolation of the support of conclusion B^* belonging to observation A^* by partially linear approximation of $\mathcal{R}_L(x)$ (the "lower function") and $\mathcal{R}_U(x)$ (the "upper function")

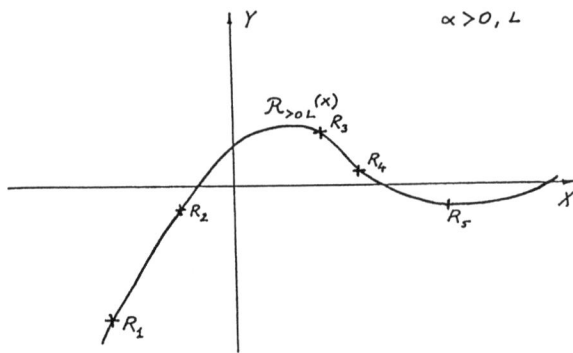

Figure 8: Smooth approximation of $supp(\mathcal{R}(x))$ by extended interpolation/extrapolation method

weighted by the reciprocal distance from the observation – as a matter of course, for every $\alpha \in \Lambda$ separately. As the location of observation is usually not exactly in the middle of the rule base, the notion of (fuzzy) distance must be extended to *signed distances*, and rules should be interpreted with alternating signs. In this way, it is possible to *fit a smooth polynomial style curve* to an arbitrary number of points – representing the α-cuts of the rules. Such a curve is often a much better approximation of $\mathcal{R}_{\alpha\ L/U}(x)$ than the piecewise linear way. An example is depicted on Figure 8.

In this manner, the approximation of the fuzzy function can be determined even for the domain of X outside of the rules, i.e. real *extrapolation* can be done. The extended form of (4) is

$$inf_{\prec}\{B_\alpha^*\} = \frac{\sum_{i=1}^r \frac{1}{(-1)^{i-1}\tilde{d}_{\alpha L}(A_{i,\alpha},A_\alpha^*)} inf_{\prec}\{B_{i,\alpha}\}}{\sum_{i=1}^r \frac{1}{(-1)^{i-1}\tilde{d}_{\alpha L}(A_\alpha^*,A_{i,\alpha})}}$$

$$sup_{\prec}\{B_\alpha^*\} = \frac{\sum_{i=1}^r \frac{1}{(-1)^{i-1}\tilde{d}_{\alpha U}(A_{i,\alpha},A_\alpha^*)} sup_{\prec}\{B_{i,\alpha}\}}{\sum_{i=1}^r \frac{1}{(-1)^{i-1}\tilde{d}_{\alpha U}(A_\alpha^*,A_{i,\alpha})}} \quad (5)$$

6 Inconsistent rule bases

The above techniques are usable if the rules in the base are essentially consistent. If there is a single expert specifying the rules of controlling an industrial process, even if the rules are slightly away from the ideal $\mathcal{R}(x)$,

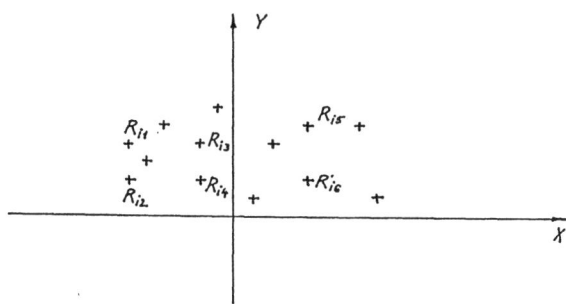

Figure 9: Inconsistent rule base with contradicting rules

the function obtained by (5) will be near to it. In many cases however, the source of information is not homogeneous, which might result into an *internal inconsistency* of the rules. Beside multiple sources of rules distortion of the information can happen, too. A common feature of such rule bases is that there is some difficulty in detecting the real tendency of the fuzzy function between X and Y.

Let us consider the rule base on Figure 9. The rough tendency of the fuzzy function is $\mathcal{R}(x) = y_0$ where y_0 is somewhere between the extremal positions of the individual rules. What happens if a piecewise linear or a general type smooth interpolation is applied? In the first case a "curve" with many very steep slopes (alternatingly negative and positive) will be found, in the second, a polynomial style curve with too large "amplitudes", very likely assuming values outside of the range of Y! On Figure 9, such a situation is illustrated: there are several rules which are even crisply contradicting: such are R_{i1} and R_{i2}, R_{i3} and R_{i4}, further on R_{i5} and R_{i6}. Here, linear interpolation is ambiguous, and the general interpolation method according to (5) results in a curve with singularities (perpendicular tangents). In the next, some methods are referred to which are able to cope with such problems, too.

In such a case, there are several strategies to use.

1. Filtering by credibility

If there are only a few rules wich contradict the majority of the rule base, they might have been included by some mistake, or the information contained in them might be noisy. If there is some additional information available, often such rules can be filtered out, and left completely out of consideration for the calculation of B^*. In other cases, just by recognizing

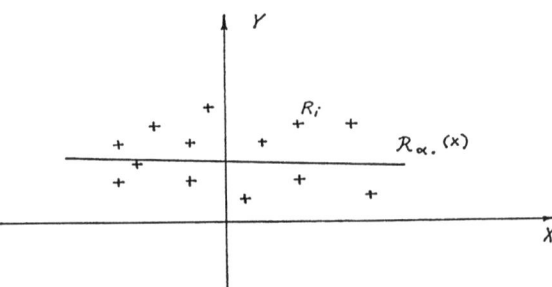

Figure 10: Simple linear regression as the estimation of $\mathcal{R}(x)$ for the contradictory rule base on the previous figure

that those rules are very much outside of the tendency of the rest of the rule base, they can be eliminated. In this paper, we do not deal with this technique, as it is application dependent and sometimes rather obvious.

2. *Approximation of $\mathcal{R}(x)$ by averaging the rules*

This method can be succesfully used if the overall tendency of the rule base i.e. $\mathcal{R}(x)$ is known. We understand this as e.g. knowing that the function is linear, piecewise linear or has a parabolic tendency over a given interval of X, etc. There are several ways of finding the best fitting curve of a certain type to a set of points. We suggest here the *least square error* approach which is well know in regression theory. The advantage of applying least square estimates is that a huge apparatus is available for that, well known results include multidimensional approximation, and a regression curve or line is very attractive from the point of view of subjective evaluation of the method. For the summary of this techniques we refer e.g. to [10]. The calculation of nonlinear regression curves is rather time consuming. Linear regression is however a convenient technique to use, and it is applicable in multiple dimensions (compound X), too. Figure 10 depicts the least square error estimation of $\mathcal{R}(x)$ by a single line, for the rule base of Figure 9, in the form $y = ax + b$, rather conforming with the expected almost horizontal position.

The equation of the best fitting straight line is given by

$$y = ax + b = \frac{\sum x_i y_i - \sum x_i \sum y_i / r}{\sum x_i^2 - \sum x_i^2 / r} x + \left(\sum y_i / r - a \sum x_i / r\right) \quad (6)$$

for one dimensional X.

It is more complicated to treat compound variables. If X has m and Y has n components, the least square regression will result a $m \times n$ dimensional hyperplane. The problem can be always decomposed into n $m+1$-dimensional problems where Y_i is approximated by $\sum_{j=1}^{m} a_{ij}x_j + b_i$. So it is sufficient to examine the case with compound X but simple Y.

The solution of this problem is given by

$$\boldsymbol{a} = [a_i] \text{ and } a = \sum_i y_i/r - \boldsymbol{a}^T [\sum_j x_{ij}] \text{ where}$$

$$\boldsymbol{a} = ([x_{ij} - \sum_j x_{ij}/r]^T [x_{ij} - \sum_j x_{ij}/r])^{-1} [x_{ij} - \sum_j x_{ij}/r]^T [y_i - \sum_i y_i/r] \quad (7)$$

$i = 1...r$, $j = 1...m$, [ ] stands for indicating a matrix, ^{T} is the transposed.

It is clear that the straight regression line and hyperplane give a rough approximation of $\mathcal{R}(x)$, except if it has a real linear tendency like the rule base on Figure 10 seems to have. In order to obtain a better approximation, it is reasonable to apply a *piecewise linear function* as the estimate of $\mathcal{R}$ and to calculate the regression line in the form $y = a(A^*)x + b(A^*)$, preferably always for a given environment of the observation: a 'window' around the respective value of A^*. Then, the best fitting straight line is obtained only for a restricted area, such that linearity can be assumed. If the window is not too large, this leads to a rather good piecewise linear approximation. Figure 11 demonstrates this for a rule base where the rules seem to fit to approximate straight lines, one having a positive slope, the other a negative one. Of course, this rule base seems to be rather sparse, so the credibility of such an approximation is doubtful. It is possible to fit the size of the "window" to the local denseness of the rule base, so approximation can be rather fine.

Let us compare the window-regression technique with the previous interpolation/extrapolation method. While in the latter, it is sufficient to calculate the approximation curve (maybe piecewise linear) once before starting the inference/control algorithm based on the rules, it seems that because of shifting the window around the observation, it is necessary to calculate a new equation for y every time when we have a new observation. This would be rather painful, especially in the compound case as the matrix inversion takes a very long runtime. Any method where for every observation a new approximation of $\mathcal{R}$ must be calculated is not competitive in computational complexity with techniques where the approximation can be

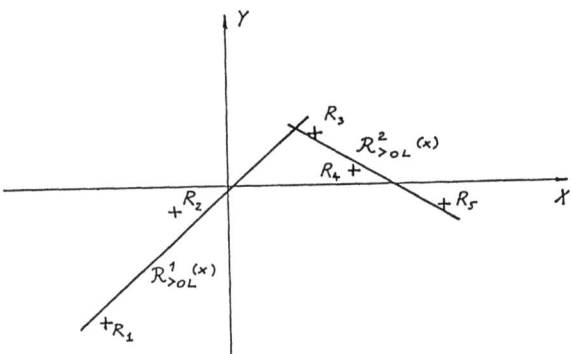

Figure 11: Piecewise linear approximation of $\mathcal{R}(x)$ in a rule base with two different approximately linear parts

established "forever", before the actual control starts (like in the case of piecewise linear interpolation with a fixed rule base). Luckily enough, in the case of simple variables and a rule base with r rules it is sufficient to calculate maximally $2r$ different regression lines for any fixed set of points and even in the compound variable case the space $X \times Y_i$ can be divided into maximally $2r^m$ areas where the regression hyperplane would be different. (For given α and lower/upper points).

Significant disadvantage of this technique is that the function obtained is not continuous: it is a line or plane with discontinuity points or edges, at the boundary of a new rule entering the shifted window. So the approximated conclusion might change abruptly when the observation is only slightly different. A solution of this problem is presented by the application of the combination of *fuzzy window and linear regression technique*, where the above method is modified so that the environment of every observation has fuzzy rather than crisp boundaries. So the abrupt appearance of a new rule in the window when the observation is moved slightly is eliminated completely: every rule appearing in the window is weighted by the membership value attached to the location of that rule – depending on the location of the observation – this weight is however very small if the window is defined by a membership function with not very steep boundaries. For this purpose, e.g. a trapezoidal window is rather suitable. Figure 12 illustrates both types of "window": the upper one is a crisp, the lower one is a trapezoidal environment. For the same rule environment and the same observation, different

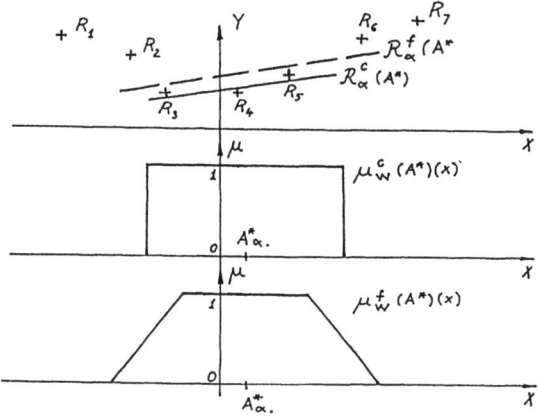

Figure 12: Crisp and fuzzy window around observation A^*. The resulting local approximation of $\mathcal{R}(x)$ is different

regression line will be calculated if the rules lying farther from the observation, have a less than 1 weight. (In the example, R_2 and R_6 are completely ignored with the crisp window, but considered with the fuzzy window – to a certain degree. On the other hand, R_3 has a less weight in the second case.)

In the case of a fuzzy window, i.e. weights in $(0, 1]$ attached to the points representing the α-cuts of rules, the least square method according to (6) and (7) must be generalized. It must be remarked here, that the gain in the smoothness and continuity of the approximation function costs considerable computational time. Because of the introduction of the fuzzy (continuous membership function) window, it is not possible to extend the statement concerning the finiteness of the number of possible regression lines. The regression line calculated in terms of the observation is continuously changing, in fact, it gives only the tangent of the real approximation function.

The next equation presents the solution of the general fuzzy regression, where points can be weighted by arbitrary membership degrees. Suppose that we have points (x_i, y_i) $(i = 1...r)$ and each has the membership degree μ_i. The straight line with least square sum of difference is then

$$y = ax + b = \frac{\sum \mu_i x_i (y_i - \frac{\sum \mu_i y_i}{\sum \mu_i})}{\sum \mu_i x_i (x_i - \frac{\sum \mu_i x_i}{\sum \mu_i})} x + \left(\frac{\sum \mu_i y_i}{\sum \mu_i} - \frac{\sum \mu_i x_i}{\sum \mu_i} a\right) \quad (8)$$

The approximation of $\mathcal{R}(x)$ is depicted on Figure 13 for the same rule

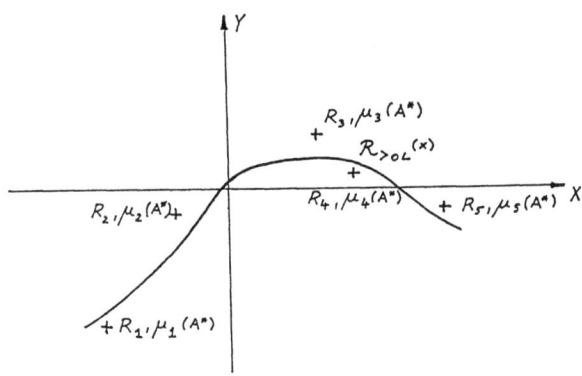

Figure 13: Approximation of fuzzy function represented by rule base $R = \{R_1, ..., R_5\}$ with trapezoidal fuzzy window and linear regression

base as on Figure 10. It is possible to extend the above result for compound variable cases. Instead of giving the rather complicated equation we just indicate that the mean values $\sum_j x_{ij}/r$ and $\sum_i y_i/r$ must be replaced by $\sum_j \mu_j x_{ij}/\sum_i \mu_i$ and $\sum_i \mu_i y_i/\sum_i \mu_i$, resp., further on, in all the sums x_i is replaced by $\mu_i x_i$. It is a rather serious problem here that computational complexity is high, in every step of inference the inversion of several $r \times k_1$ dimensional matrices is to be done – depending on the cardinality of level sets at least 3 or 4 of them. It is still an open question if the approximative curve can be calculated directly, before the actual control starts. If the function of the fuzzy window can be expressed analytically, the integral of (8) might be determinable. In other cases, a numerical solution might reduce the high computational time.

Some more details on these topics can be found in [11, 12].

3. *Inference with more than one simultaneous tendencies in the rule base*
A completely different idea is the recognition of simultaneous tendencies in the inconsistent rule base, as e.g. when at least two experts with dissonant knowledge are represented in R. As in contrary to the averaging technique where the possible "mean opinion" is attempted to be found, here the aim is the separation of the main streams in the rule base, i.e. the determination of several rule clusters representing different functions (although similar in tendency). Returning to the rule base on Figure 9, and reexamining the rules, we find that two simultaneous tendencies are present, one is approximately $\mathcal{R}(x) = y_1$ while the other one is $\mathcal{R}(x) = y_2$, where $y_1 \neq y_2$ and y_0 in

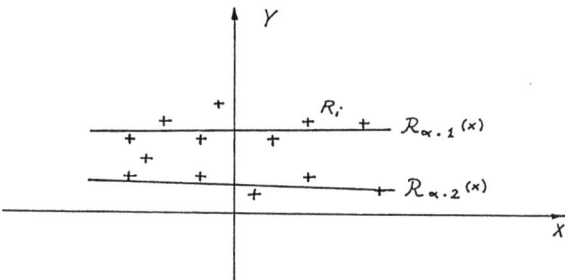

Figure 14: Two simultaneous tendencies in an inconsistent rule base, detected by fuzzy Hough transform

the previous linear regression approach lies somewhere between y_1 and y_2.

Such simultaneous tendencies – various straight lines fitting several clusters of points – can be detected by computer vision methods, where edge detection represents a similar problem. A very attractive method is the application of Hough-transform [13] which however has the disadvantage of finding only exactly fitting lines in the original form. The technique of fuzzy Hough transform proposed in [14] eliminates this difficulty and offers a means to find parallel approximate tendencies in a rule base, too – provided that such tendencies are present and consonant for every $\alpha \in \Lambda$. The result of such an analysis of the same rule base can be found on Figure 14, here the degree of fuzzifying the points is not very high - otherwise the two tendencies merge into a single line, similar to the one obtained when using regression for the total rule base (however, in this case, very singularly located points do not influence the approximated function!).

Research in this direction is going on.

7 Large state spaces with hierarchical rule systems

Finally, the problem of structured control in state spaces with a very large number of variables or otherwise very complex systems is mentioned. In this context, structured control means that a hierarchical rule base is applied. Very promising results have been published recently in this area (see e.g. [15]), however still many theoretical and practical questions have to be answered. As an example for a hierarchically structured fuzzy control system, let us have a state space with only two input variables, and let us construct a simple rule system:

Level 1:
"If X is D_1 then R_1"
"If X is D_2 then R_2"
"If X is D_3 then R_3"

Level 2 (R_1, R_2 and R_3):

R_1:
"If X_1 is positive small then Y is big"
"If X_1 is negative small then Y is medium"
"If X_1 is negative big then Y is small"

R_2:
"If X_1 is not big and X_2 is small then Y is medium"
"If X_1 is big or X_2 is not small then Y is big"

R_3:
"If X_2 is positive small then Y is medium"
"If X_2 is negative not small then Y is small"

In this system it can be noticed that some rules do not refer to both variables in the antecedent part. Inside D_1 only X_1 determines the conclusion, X_2 can be completely left out of consideration. In D_2 both variables must be considered, while in D_3 only X_2 dominates the situation. In a real applicational problem, the number of inputs might be as high as several hundred, while in the practice, not more than a few dozen can be treated effectively. By applying this heterogenous style rule base, serious reduction of computational time can be achieved (cf. [16]). In many practical systems, depending on the location of the observation, a few input variables might determine the system completely. When moving to another domain in X, the dominant combination of variables might change, the critical case is the borderline area where both (or several) groups have their effect.

For domains D_i in the above rule system, see Figure 15. The rule system is hierarchically structured in the sense that for a given observation A^* first its position in the state space must be found. If the figure refers to the supports of fuzzy domains D_i, and the support of the observation is located e.g. so that $proj_{X_1}(supp(A^*)) > x_{12}$ and $proj_{X_2}(supp(A^*)) > x_{22}$ then A^* overlaps only with D_2. That is, the conclusion of step one in the inference is "R_2". Then again, the rules in R_2 must be applied in such a way as it is done in a non hierarchical control system. If however, A^* is such that its support is e.g.

$$inf_{X_1}\{supp(A^*)\} < x_{11}$$
$$sup_{X_1}\{supp(A^*)\} > x_{12}$$
$$inf_{X_2}\{supp(A^*)\} < x_{21}$$
$$sup_{X_2}\{supp(A^*)\} > x_{22}$$

then all three rule subsystems must be applied simultaneously. This raises the new question, how to approximate $\mathcal{R}(x)$ in such an inhomogenous case where R_1 might partially contradict to R_2, etc.? Also, it is interesting to investigate what happens if D_i are fuzzy domains themselves. In such a case, for a given observation, every rule subsystem might be valid to a certain degree in $(0, 1]$, a kind of internal contradiction which must be resolved by some technique similar to those mentioned in connection with simple (one level) rule bases. This will often occur in the borderline areas among several fuzzy domains D_i. The final control action will be a weighted combination of various conclusions originating from various rule subsystems partially or fully overlapping with the observation. This kind of evaluation of the observation is more advantageous from the point of view of computational complexity than a combined rule system with many input variables occuring in all the rules.

An even more interesting question is, what happens if the rule bases are partially of the Takagi-Sugeno type [17] where rules have the form

"*If X is A_i then $y = f(x)$*"

Some related problems will be also treated in [16].

References

[1] L. A. Zadeh: Outline of a new approach to the analysis of complex systems and decision processes, *IEEE Trans. System, Man and Cybernetics (1973)*. pp. 28-44.

[2] E. H. Mamdani and S. Assilian: An experiment in linguistic synthesis with a fuzzy logic controller. In: *E.H. Mamdani and B. R. Gaines (eds.): Fuzzy Reasoning and Its Applications.* Academic Press, London, 1981. pp. 311-323. (Originally published in 1975 in *Int. J. of Man-Machine Studies* in 1975.)

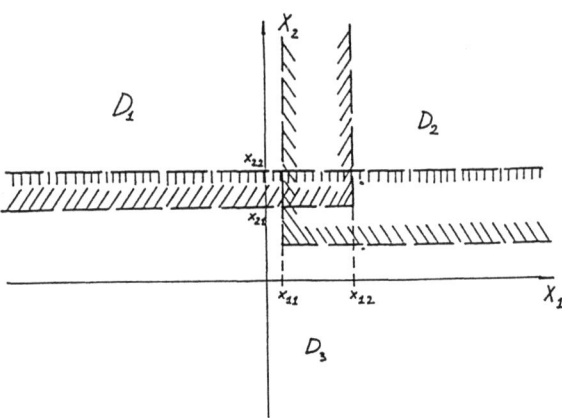

Figure 15: Rule subsystem validity domains in a compound input variable space

[3] D. G. Burkhardt and P. P. Bonissone: Automated fuzzy knowledge base generation and tuning. *IEEE Int. Conf. on Fuzzy Systems*. San Diego, Calif., 1992, pp. 179-196.

[4] L. T. Kóczy: On the computational complexity of rule based fuzzy inference. *NAFIPS – '91*. Columbia, Missouri, 1991. pp. 87-91.

[5] D. Dubois and H. Prade: Gradual inference rules in approximate reasoning. To appear in *Information Sciences*, 1992.

[6] I. B. Türkşen and Z. Zhong: An approximate analogical reasoning approach based on similarity measures. *IEEE Trans. on Systems, Man and Cybernetics*, 1988. pp. 1049-1056.

[7] L. T. Kóczy and Á. Juhász: Fuzzy rule interpolation and the RULEINT program. In: *L. T. Kóczy and K. Hirota (eds.): Joint Hungarian – Japanese Symposium on Fuzzy Systems and Applications*. Budapest, 1991. pp. 91-94.

[8] L. T. Kóczy, K. Hirota and Á. Juhász: Interpolation of 2 and 2k rules in fuzzy reasoning. *Fuzzy Engineering Toward Human Friendly Systems I*. Yokohama, 1991. pp. 206-217.

[9] L. T. Kóczy and K. Hirota: Reasoning by analogy with fuzzy rules. *IEEE International Conference on Fuzzy Systems*. San Diego, 1992. pp. 263-270.

[10] N. R. Draper and H. Smith: Applied regression analysis. 2nd ed. Wiley, New York, 1981.

[11] L. T. Kóczy: Inference in fuzzy rule bases with conflicting evidence. Submitted to *NAFIPS '92*. Puerto Vallarta, 1992.

[12] L. T. Kóczy: Reasoning by analogy with sparse fuzzy rule bases. *Korea Fuzzy Mathematics and Systems Society Spring Conference*. Seoul, 1992. pp. 5-16.

[13] R. O. Duda and P. E. Hart: Use of the Hough transformation to detect lines and curves in pictures. *Communications of the ACM 15*, 1972. pp. 11-15.

[14] J. H. Han, L. T. Kóczy and T. Poston: Fuzzy Hough transform for sparse data. Manuscript under submission, 1992.

[15] M. Sugeno: Demonstration of unmanned helicopter control at *Fuzzy Engineering Toward Human Friendly Systems I*. Yokohama, 1991.

[16] L. T. Kóczy and K. Hirota: Manuscript in preparation for *FUZZ-IEEE '93*, on the interpolation of rule bases in hierarchically structured fuzzy control systems

[17] T. Takagi and M. Sugeno: Fuzzy identification of sytems and its applications to modeling and control. *IEEE Trans. on Systems, Man and Cybernetics 15*, 1985. pp. 116-132.

FUZZY SETS AND POSSIBILITY THEORY :
SOME APPLICATIONS TO INFERENCE AND DECISION PROCESSES

Didier DUBOIS, Henri PRADE
Institut de Recherche en Informatique de Toulouse (I.R.I.T.) – C.N.R.S.
Université Paul Sabatier, 118 route de Narbonne, 31062 Toulouse Cedex, France

Abstract : Fuzzy sets model extensions of vague predicates, as well as flexible constraints or objectives. The concept of a possibility distribution enables to represent imprecise, uncertain, or vague states of information. Thus fuzzy sets and possibility theory together offer a framework for approximate and uncertain reasoning. The paper illustrates these two dimensions of fuzzy set-based methods, namely the handling of soft requirements and the modeling of imprecise and uncertain information, in inference and decision processes. More particularly, approximate reasoning in fuzzy control, application of fuzzy arithmetics to qualitative reasoning, and application of twofold fuzzy sets to multiple criteria aggregation are discussed.

1. Introduction

When we encounter a statement like "John is small", or more generally "the value of variable X is small", it may have two different intended meanings, both of them taking advantage of the vagueness pervading a word such as "small".

On the one hand the statement can take place in a situation where the value of X is precisely known and we estimate the extent to which this value is compatible with the label "small" (whose meaning obviously depends on the context). In this case we are interested in the gradual, flexible nature of the specification stated by "X is small" which may express a constraint, or a criterion (for instance, one looks for somebody who should be small and we estimate to what extent John satisfies this requirement, e.g. John (whose height is known to be 1.65 m) can be considered as "small" at the degree 0.8 in a considered context). More particularly the concept of a fuzzy set deals with the representation of classes whose boundaries are not quite determined, by means of characteristic functions taking values in the interval [0,1]. In a fuzzy class, some elements that are considered as "marginal" or "less acceptable", are given a degree of membership that is intermediary between 0 and 1.

On the other hand, "X is small" may mean as well "all we know about the value of X, is that X is small" (without knowing the value of X precisely in this case). This corresponds to a situation of incomplete information (pervaded with imprecision and uncertainty), a situation in which we can only order the possible values of X according to their level of plausibility or possibility. When a fuzzy set is used in order to represent what is known about the value of a variable, the degree attached to a value expresses the level of possibility that this value is indeed the value of the variable. However in case of incomplete information, several values may have a degree of possibility equal to 1. Moreover, we shall be all the more certain that X takes its value in some subset A as all the values outside A have a smaller degree of possibility.

Fuzzy sets (Zadeh, 1965) and possibility theory (Zadeh, 1978), offer a unified framework for taking into account the gradual or flexible nature of specifications, and the representation of incomplete information. The paper, which partially borrows from (Dubois and Prade, 1992b), illustrates this view ; it is organized in the following way. Section 2 discusses the representation of uncertain and imprecise information in possibility theory. Section 3 recalls the general principles of approximate reasoning. Section 4 presents different kinds of fuzzy rule-based inferences with a fuzzy control perspective. Section 5 suggests some applications of fuzzy arithmetics to qualitative reasoning. Section 6 discusses multiple criteria aggregation when we distinguish between values which are more or less certainly acceptable and values which are more or less certainly *not* acceptable according to a criterium.

2. Represention of Imprecise and Uncertain Information

The theory of approximate reasoning, as formulated by Zadeh (1979), can be viewed as an application of possibility theory (Zadeh, 1978 ; Dubois and Prade, 1988). It is essentially a methodology for representing some available information (especially when it is vague or incomplete) in terms of possibility distributions and for deducing from this information what can be said about the values of variables of interest. Hence a proper understanding of approximate and uncertain reasoning presupposes that the basic principles of possibility theory are well-understood.

Let U be a set that represents the range of a variable x. Usually x stands for *the* unknown value taken by some single-valued attribute applied to an object under consideration. For instance x refers to the age of a man named Peter. A possibility distribution π_x on U is a mapping from U to the unit interval [0,1] attached to the single-valued variable x. The function π_x represents a flexible restriction which constrains the possible values of x according to the available information, with the following conventions for interpreting the possibility scale [0,1] :

$\pi_x(u) = 0$ means that x = u is definitely impossible,
$\pi_x(u) = 1$ means that absolutely nothing prevents that x = u

Note that $\pi_x(u) = 1$ and $\pi_x(u') = 1$, with u' ≠ u, are allowed at the same time. Flexibility in this constraint about the possible values of x is modelled by letting $\pi_x(u)$ between 0 and 1 for some values u. The quantity $\pi_x(u)$ thus represents the degree of possibility of the assignment x = u, some values u being more possible than others, according to what is known. In other words, if we consider the logical propositions p_u whose semantic contents is 'x = u', π_x encodes a *preference* relation among these propositions describing possible states of the world : the world where Peter is 20, the world where Peter is 35 and so on in the above example ; if we prefer, the relation $\prec$ defined by $u \prec u' \Leftrightarrow \pi_x(u) < \pi_x(u')$ expresses that u' is strictly preferred to u as a possible candidate for the value of x according to the available information, among more or less possible (or allowed) values, i.e. among the u's such that $\pi_x(u) > 0$. Clearly, if U exhaustively contains all the values which may be thought of for x, at least one of the elements of U should be fully possible as a value of x, so that $\exists u, \pi_x(u) = 1$ (*normalization*). To summarize, a possibility distribution π_x encodes a preference relation about the possible values of the variable x, acording to the available knowledge.

Given a possibility distribution π_x, several set-valued functions of interest can be defined in order to estimate the tendency of x to belong to a given subset A of U. Two of these set-functions are well-known and widely used, namely

– the so-called measure of possibility (Zadeh, 1978)

$$\Pi(A) = \sup_{u \in A} \pi_x(u) \qquad (1)$$

which estimates the *consistency* of the statement 'x ∈ A' with what we know about the possible values of x ; hence we look for the element(s) of A having the greatest possibility degree according to π_x

— the dual measure of necessity,

$$N(A) = 1 - \Pi(\bar{A}) \tag{2}$$
i.e., $\quad N(A) = \inf_{u \notin A} 1 - \pi_X(u) \tag{3}$

which estimates to what extent each value outside A, i.e. in the complement $\bar{A}$ of A, has a low degree of possibility, thus the values with the higher degrees of possibility should be among the elements of A, which makes us somewhat *certain* that indeed x belongs to A. The duality relation (2) between Π and N expresses that A is all the more certain as $\bar{A}$ is less consistent with the available knowledge. It assumes that π_X is normalized. The normalization of π_X entails $\max(\Pi(A), \Pi(\bar{A})) = 1$, and then $N(A) > 0$ entails $\Pi(A) = 1$, i.e. A should be completely consistent with what is known before being somewhat certain.

A measure of 'guaranteed possibility' (Dubois and Prade, 1992a) can be also introduced

$$\Delta(A) = \inf_{u \in A} \pi_X(u) \tag{4}$$

which estimates to what extent *all* the values in A are actually possible for x according to what is known, i.e. any value in A is at least possible for x at the degree $\Delta(A)$. Clearly Δ is a stronger measure than Π, i.e. $\Delta \leq \Pi$, since Π only estimates the existence of at least one value in A compatible with the available knowledge, while the evaluation provided by Δ concerns *all* the values in A. Note also that Δ and N are unrelated, since N(A) estimates the certainty that the value of x is in A by checking the impossibility of all the values out of A, while $\Delta(A) > 0$ considers the possibility of all the values in A. Δ is a monotonically decreasing set function (in the wide sense) with respect to set inclusion; it contrasts with Π and N which are monotonically increasing.

A possibility distribution is not always specified as such, but often through the qualification of subsets of U. Let A be an *ordinary* subset of U. Two kinds of specification exist, namely

i) we know that it is (completely) *certain* that the value of x lies in A (for short "A is certain for x"). It means that any value outside A is (completely) impossible, i.e. $\forall u \notin A, \pi_X(u) = 0$ and π_X is unspecified over A, or if we prefer

"A is certain for x" is translated by $\forall u \in U, \pi_X(u) \leq \mu_A(u) \tag{5}$

where μ_A is the $\{0,1\}$-valued characteristic function of A;

ii) A is a (completely) possible range for x (for short "A is possible for x"). It means that $\forall u \in A, \pi_x(u) = 1$ and π_x remains unspecified outside of A, or equivalently that

"A is possible for x" is translated by $\forall u \in U, \mu_A(u) \leq \pi_x(u)$. (6)

In the first kind of specification we express our certainty, delimiting a subset A which for sure contains the value of x ; the larger A, the more imprecise we remain about the value of x, the smaller A, the better the information. On the contrary, in the second kind of specification, we rather express our uncertainty, stating a range A of values which are indeed possible for x without rejecting any value outside of A ; in some sense we are recognizing the extent of our ignorance with respect to the precise value of x.

Let us now consider the cases of qualification where the possibility or the certainty is not complete but corresponds to an intermediary level α in the scale [0,1]. It leads to the two following generalizations of (5) and (6)

i) The statement "it is certain at least at the degree α that the value of x is in A", will be interpreted as "any value outside A is at most possible at the complementary degree, namely $1 - \alpha$", i.e. $\forall u \notin A, \pi_x(u) \leq 1 - \alpha$, which leads to

"A is α-certain for x" is translated by $\forall u \in U, \pi_x(u) \leq \max(\mu_A(u), 1 - \alpha)$ (7)

It can be checked that this is equivalent to $N(A) \geq \alpha$. Note that for $\alpha = 1$, (5) is recovered. When α decreases from 1 to 0, our knowledge evolves from complete certainty in A to acknowledged ignorance about x.

ii) The statement "A is a possible range for x at least at the degree α" will be understood as $\forall u \in A, \pi_x(u) \geq \alpha$, which leads to

"A is α-possible for x" is translated by $\forall u \in U, \min(\mu_A(u), \alpha) \leq \pi_x(u)$. (8)

It can be checked that this is equivalent to $\Delta(A) \geq \alpha$. Note that for $\alpha = 1$, (6) is recovered. When α decreases from 1 to 0, our knowledge evolves from the certainty that A is the minimal range of our ignorance to a total lack of information whatsoever.

The above representations (7) and (8) can be generalized to the case where A is a *fuzzy* set, namely

i) $\forall u \in U, \pi_x(u) \leq \max(\mu_A(u), 1 - \alpha)$ (α-certainty qualification)

ii) $\forall u \in U, \pi_x(u) \geq \min(\mu_A(u), \alpha)$ (α-possibility qualification).

It is still equivalent to $N(A) \geq \alpha$ and $\Delta(A) \geq \alpha$ provided we use the following appropriate extensions of N and Δ when A is fuzzy

$$N(A) = \inf_{u \in U} (1 - \mu_A(u)) \to (1 - \pi_x(u)) \qquad (9)$$
$$\Delta(A) = \inf_{u \in U} \mu_A(u) \to \pi_x(u) \qquad (10)$$

where $\to$ is the so-called Gödel's implication $r = s \to t = \begin{cases} 1 & \text{if } x \leq t \\ t & s > t \end{cases}$. Indeed we have the equivalences $\max(s, 1 - r) \geq t \Leftrightarrow r \leq (1 - s) \to (1 - t)$ and $\min(s, r) \leq t \Leftrightarrow r \leq s \to t$.

3 - General Principles of Approximate Reasoning

In the previous section we have seen how the representation of certainty and possibility qualifications lead to inequality constraints on possibility distributions. In case of certainty qualification, the inequality is of the type $\pi_x \leq \mu_A$ where μ_A may be any [0,1]-valued (normalized) membership function; it includes the case where μ_A is of the form $\mu_A = \max(\mu_F, 1 - \alpha)$, i.e. "F is α-certain for x" is equivalent to "A is certain for x" (where A is then necessarily a fuzzy set). The *principle of minimal specificity* stipulates that each value u in the domain U of a variable x (in the general case x is a vector and U a Cartesian product) should receive the largest degree of possibility which is in agreement with the constraint(s); i.e. it leads to take $\forall u \in U, \pi_x(u) = \mu_A(u)$ in the above case. Indeed choosing a particular π such that $\pi < \mu_A$ in order to represent our knowledge about x would be arbitrarily too precise. The principle of minimal specificity has a role in possibility theory similar to the use of the maximal entropy principle in probability theory. The notion of specificity of a possibility distribution has been introduced by Yager (1983). The absence of constraint corresponds to the situation of complete ignorance modelled by $\forall u \in U, \pi_x(u) = 1$. At the opposite, complete knowledge corresponds to a maximal constraint, such that $\exists u_0, A = \{u_0\}$, i.e. $\pi_x(u_0) = 1$ and $\pi_x(u) = 0, \forall u \neq u_0$. In that sense any possibility distribution π_x is provisional in nature and likely to be improved by further information, when the available one is not complete. When several pieces of information constraining the possible values of x become available, i.e. $\pi_x \leq \mu_{A_i}$, $i = 1, n$, the principle of minimum specificity leads to assume

$$\pi_x = \min_{i=1,\ldots,n} \mu_{A_i} \qquad (11)$$

as a straightforward consequence of the inequalities. Note that the complete reliability of the sources entails that $\sup_u \min_{i=1,\ldots,n} \mu_{A_i}(u) = 1$. Otherwise, the sources are conflicting as soon as $\pi_x(u) < 1$, $\forall u$. Hence subnormalization corresponds to *partial inconsistency*. Clearly if we have two constraints $\pi_x \leq \mu_{A_1}$ and $\pi_x \leq \mu_{A_2}$, such that $\mu_{A_1} \leq \mu_{A_2}$, (i.e., we have the fuzzy set inclusion $A_1 \subseteq A_2$) the second piece of information is redundant and A_1 is said to be more specific than A_2.

Obviously a different information principle needs to be applied in case of possibility qualification where we deal with inequalities of the type $\pi_x \geq \mu_A$. Here μ_A does not need to be normalized, indeed μ_A may be of the form $\mu_A = \min(\alpha, \mu_F)$. A *principle of maximal specificity* can be used in this case to limit the scope of the possible values for x to the values which are indeed known as (somewhat) possible. While the principle of minimal specificity embodies the commonsense claim that anything that is not impossible should not be ruled out, the converse principle expresses that anything which is not established as being possible can be neglected. The role of information principles in possibility and other non-standard theories of uncertainty is particularly stressed in the book by Klir and Folger (1988).

Let us now assume that there are two variables x and y ranging on U and V respectively, with possibility distributions π_x and π_y respectively. The principle of minimal specificity leads us to define the joint possibility distribution $\pi_{x,y}$ as the *combination*

$$\pi_{x,y} = \min(\pi_x, \pi_y) \tag{12}$$

as a consequence of the inequalities $\pi_{x,y}(u,v) \leq \pi_x(u)$, $\forall v$ and $\pi_{x,y}(u,v) \leq \pi_y(v)$, $\forall u$, which express that the possibility that $x = u$ *and* $y = v$ is necessarily upper-bounded by the possibility that $x = u$ (nothing being said about y) on the one hand and the possibility that $y = v$ on the other hand. Note that (12) is in agreement with (11). Indeed the possibility distribution π_x on U (resp. π_y on V) can be extended to $U \times V$ into a possibility distribution $\pi^1_{x,y}$ (resp. $\pi^2_{x,y}$) by stating that $\forall u$, $\forall v$, $\pi^1_{x,y}(u,v) = \pi_x(u)$ (resp. $\pi^2_{x,y}(u,v) = \pi_y(v)$), since nothing is said about y (resp. x) ; this is called *cylindrical extension*. Then $\pi_{x,y} = \min(\pi^1_{x,y}, \pi^2_{x,y})$ from (11). As it can be seen it is equivalent to first apply the minimal specificity principle for each possibility distribution representing a piece of information and then to combine by (12), or to first combine the constraints corresponding to certainty-qualified statements and then to apply the minimal specificity principle to the result.

Conversely, given a possibility distribution $\pi_{x,y}$ restricting the possible values of a pair of variables (x,y) the induced restriction on the possible values of x (resp. y) can be obtained by the *projection* of $\pi_{x,y}$ on U (resp. V), i.e.

$$\pi_x(u) = \sup_{v \in V} \pi_{x,y}(u,v) \;;\; \pi_y(v) = \sup_{u \in U} \pi_{x,y}(u,v) \qquad (13)$$

Again this is in agreement with the principle of minimal specificity since we compute the largest possibility degrees satisfying the constraint x = u, ∀ u ∈ U (resp. y = v, ∀ v ∈ V). The combination can be straightforwardly extended to n-tuples of variables and the projection can be performed on any subset of such tuples. Combination and projection as defined above are the basis of the theory of approximate reasoning introduced by Zadeh (1979). The combination and projection operations in possibility theory play roles which are respectively similar to the construction of a joint distribution (under a stochastic independence assumption) and to the computation of a marginal distribution in probability theory. The simultaneous satisfaction of (12) and (13) supposes that x and y are logically independent (i.e. that the degree of possibility that x = u does not depend on the value of y, ∀u, and similarly exchanging x and y).

An important point related to computational issues is the fact that the results of elementary combination and projection steps can be often analytically precomputed. This is illustrated on various examples in the following section.

4. Fuzzy Rule-Based Inference

The simplest pattern of fuzzy inference is the so-called generalized modus ponens (Zadeh, 1979) which from a fact "x is A'" and a rule "if x is A then y is B", where A', A and B denote subsets which are possibly fuzzy, enables us to derive a conclusion of the form "y is B'". The (fuzzy) subset B' obviously depends on A', A and B but also on the intended meaning of the fuzzy rule "if x is A then y is B". In the following we consider certainty rules, possibility rules and gradual rules successively.

4.1. Certainty Rules

A certainty rule is a piece of knowledge of the form "the more x is A, the more certain y is B" (where A should be a fuzzy predicate). Such a rule is understood as "if X = u then y is B is $\mu_A(u)$-certain" ; then applying (7) with $\alpha = \mu_A(u)$ and the principle of minimum of specificity, it leads to model this rule by the conditional possibility distribution

$$\pi_{y|x}(v,u) = \max(\mu_B(v), 1 - \mu_A(u)).$$

From the two pieces of information "x is A'", i.e. $\pi_x = \mu_{A'}$, and the rule "the more x is A, the more certain y is B", we obtain by combination and projection

$$\pi_y(v) = \sup_u \min(\mu_{A'}(u), \max(\mu_B(v), 1 - \mu_A(u)))$$
$$= \max(\mu_B(v), 1 - N_{A'}(A)) \qquad (14)$$

where $N_{A'}(A) = \inf_u \max(\mu_A(u), 1 - \mu_{A'}(u))$ is a necessity measure of the fuzzy event A, defined from the possibility distribution $\pi = \mu_{A'}$ (not to be confused with (9)), such that $N_{A'}(A) = 1$ if and only if the support of A' is included in the core of A, i.e. $\{u, \mu_{A'}(u) > 0\} \subseteq \{u, \mu_A(u) = 1\}$. Indeed if we only know that "X is A'", i.e. $\pi_x = \mu_{A'}$, $\alpha = \mu_A(x)$ is not always equal to 1 when x ranges in the support of A if A is fuzzy, and then we cannot be fully certain that y is B. Hence (14) expresses that the conclusion "y is B" is all the more certain as the fuzzy set A' of possible values of x corresponds to elements highly compatible with A ; in other words, y is B is $N_{A'}(A)$-certain. The use of certainty rules thus lead to the propagation of certainty coefficients.

4.2. Possibility Rules for Analogical Reasoning

A possibility rule encodes a piece of knowledge of the form "the more x is A, the more possible y is B", which translates into the conditional possibility $\pi_{y|x}(v,u) = \min(\mu_B(v), \mu_A(u))$, if we interpret the rule as "if $x = u$, then y is B is $\mu_A(u)$-possible" (applying (8) and a principle of maximum specificity). Then from the fact "x is A'" and a possibility rule we deduce by combination and projection

$$\pi_y(v) = \sup_u \min(\mu_{A'}(u), \mu_B(v), \mu_A(u))$$
$$= \min(\mu_B(v), \Pi_{A'}(A)) \qquad (15)$$

where $\Pi_{A'}(A) = \sup_u \min(\mu_A(u), \mu_{A'}(u))$ is the possibility measure of the fuzzy event A based on $\pi_x = \mu_{A'}$. Hence (15) expresses that y is B is $\Pi_{A'}(A)$-possible. The values restricted by B are all the more possible for y as the fact x is A' is consistent with the condition x is A.

Note that if A' = U, i.e. we regard any value in U as possible for x, we have $\Pi_{A'}(A) = \Pi_U(A) = 1$ and y is B is fully possible. By contrast if A' is a singleton $\{u_0\}$

representing the precise fact $x = u_0$, we conclude that y is B is $\mu_A(u_0)$-possible ; in particular if $\mu_A(u_0) = 0$, we obtain $\forall\ v \in V,\ \pi_y(v) \geq 0$ (i.e. there is no constraint on π_y), provided that we did not apply the principle of maximum specificity, i.e. keeping $\pi_{y|x}(v,u) \geq \min(\mu_B(v), \mu_A(u))$ as the representation of the rule. Thus, as soon as there exists a value u which is fully possible for x (i.e. $\mu_{A'}(u) = 1$) which belongs to the core of A (i.e. $\mu_A(u) = 1$), we deduce that the values restricted by B are fully possible for y ; if all the possible values of x are outside the support of A (i.e. $\forall u,\ \mu_{A'}(u) > 0 \Rightarrow \mu_A(u) = 0$), then nothing can be said about values in V which would have a guaranteed possibility for y.

Let us now consider the case of two objects described in terms of attribute variables x and y. The indices 1 and 2 will refer to the values of these attributes for each of these objects respectively. Let us suppose we have the following information : $\pi_{x_1} = \mu_{A_1}$, $\pi_{y_1} = \mu_{B_1}$, $\pi_{x_2} = \mu_{A_2}$ and the rule "the more similar the values of x_1 and x_2, the more possible the approximate equality of the values of y_1 and y_2". Let S and T be the fuzzy relations "similar" and "approximately equal" in the above rule. Then by combination/projection we obtain for y_2 (applying (8) with $\alpha = \mu_T(v_1,v_2)$)

$$\pi_{y_2}(v_2) = \sup_{u_1,u_2,v_1} \min(\mu_{A_1}(u_1), \mu_{A_2}(u_2), \mu_{B_1}(v_1), \mu_S(u_1,u_2), \mu_T(v_1,v_2))$$
$$= \min(\mu_{B \circ T}(v_2), \sup_{u_1,u_2} \min(\mu_S(u_1,u_2), \mu_{A_1}(u_1), \mu_{A_2}(u_2)))$$
$$= \min(\mu_{B \circ T}(v_2), \Pi_{A_1 \times A_2}(S)) \tag{16}$$

where $\Pi_{A_1 \times A_2}(S)$ is the possibility measure of the fuzzy event "x_1 and x_2 are similar", given that "x_1 is A_1" and "x_2 is A_2" and (12) is used to combine μ_{A_1} and μ_{A_2}. The conclusion obtained by (16) means that the values which are restricted by B or which are approximately equal to a value in B are all the more possible for y_2 as x_1 and x_2 are indeed possibly similar. This kind of weak conclusion is in the spirit of analogical reasoning which can lead only to tentative conclusions.

4.3. Gradual Rules for Interpolative Reasoning in Fuzzy Control

A gradual rule is of the form "the more x is A, the more y is B" and is understood as "the greater the degree of membership of the value of x to the fuzzy set A and the more the value of y is possibly related to x, the greater should be the degree of membership to B for this value of y", it translates into the inequality

$$\forall\, u \in U,\, \forall\, v \in V,\, \min(\mu_A(u), \pi_{y|x}(v,u)) \le \mu_B(v). \tag{17}$$

Using again the equivalence $\min(a,t) \le b \Leftrightarrow t \le a \to b$, where $\to$ is Gödel implication, (17) can be rewritten under the form

$$\forall\, u \in U,\, \forall\, v \in V,\, \pi_{y|x}(v,u) \le \mu_A(u) \to \mu_B(v) = \begin{cases} 1 & \text{if } \mu_A(u) \le \mu_B(v) \\ \mu_B(v) & \text{if } \mu_A(u) > \mu_B(v) \end{cases}. \tag{18}$$

Let us consider a collection of gradual rules of the form "the closer x is to a_i, the closer y is to b_i" where (a_i, b_i), $i = 1, n$ are pairs of *scalar* values. These rules are supposed to specify what the value of the command variable y should be for controlling a process whose output x is precisely observed (see, e.g. Berenji, 1992). The first problem is to represent "close to a_i", by means of a fuzzy set A_i. It seems natural to assume that $\mu_{A_i}(a_{i-1}) = \mu_{A_i}(a_{i+1}) = 0$ since there are special rules adapted to the cases $x = a_{i-1}$, $x = a_{i+1}$. Moreover if $u \ne a_i$, $\mu_{A_i}(u) < 1$ for $u \in (a_{i-1}, a_{i+1})$, since information is only available for $u = a_i$. Hence A_i should be a fuzzy interval with support (a_{i-1}, a_{i+1}) and core $\{a_i\}$. Besides, by symmetry, since the closer x is to a_{i-1}, the farther it is from a_i, $\mu_{A_{i-1}}$ should decrease when μ_{A_i} increases, and $\mu_{A_i}\left(\frac{a_i + a_{i+1}}{2}\right) = \mu_{A_{i-1}}\left(\frac{a_{i-1} + a_i}{2}\right) = 0.5$. The simplest way of achieving this is to let

$$\forall\, u \in [a_{i-1}, a_i],\, \mu_{A_{i-1}}(u) + \mu_{A_i}(u) = 1$$

an example of which are triangular-shaped fuzzy sets as in Figure 1.

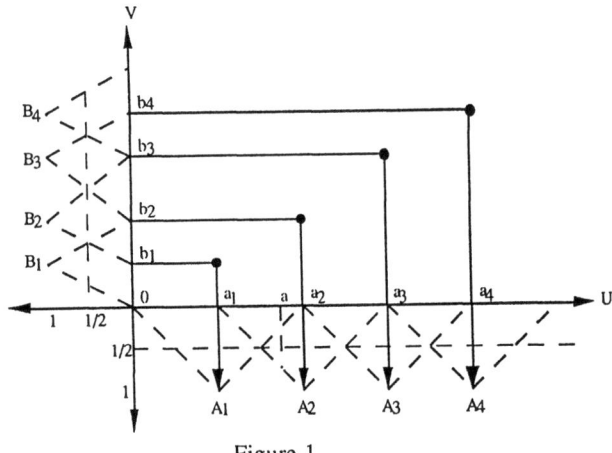

Figure 1

Clearly the conclusion parts of the rules should involve fuzzy sets B_i whose meaning is "close to b_i", with similar convention (see Figure 1). In other words, each rule is understood as "the more x is A_i, the more y is B_i", i.e. here "the closer x is to a_i, the closer y is to b_i".

In that case the fuzzy set B restricting the possible values of y, associated with the precise observation x = a, where $a_{i-1} < a < a_i$, can be easily computed from (18) by combination and projection (applying the minimum specificity principle). It gives

$$\forall u, \mu_B(v) = \sup_u \min(\mu_{\{a\}}(u), \min_{i=1,n} \mu_{A_i}(u) \to \mu_{B_i}(v)) = \min_{i=1,n} \mu_{A_i}(a) \to \mu_{B_i}(v)$$

i.e.

$$B = (\alpha_{i-1} \to B_{i-1}) \cap (\alpha_i \to B_i) \qquad (19)$$

where $\alpha \to B$ symbolically denotes the fuzzy set of membership $\mu_{\alpha \to B} = \alpha \to \mu_B$, $\to$ is Gödel implication and $\alpha_{i-1} = \mu_{A_{i-1}}(a)$, $\alpha_i = \mu_{A_i}(a)$ with $\alpha_{i-1} + \alpha_i = 1$. Then it can be easily proved (without the assumption of triangles) that there exists a unique value y = b such that $\mu_B(b) = 1$, which exactly corresponds to the result of a linear interpolation, i.e. we have $b = \alpha_{i-1} \cdot b_{i-1} + \alpha_i \cdot b_i$. It is a theoretical justification for Sugeno and Nishida (1985)'s inference method where the conclusion parts of the rules are precise values b_i and where a linear interpolation is performed on the basis of the degrees of matching $\alpha_i = \mu_{A_i}(a)$. Hence reasoning with gradual rules does model interpolation, linear interpolation being retrieved as a particular case.

This contrasts with the result obtained with Mamdani (1977)'s method for a precise observation x = a and a collection of fuzzy rules modelled in terms of what is called here "possibility rules", namely

$$\forall v, \mu_B(v) = \max_{i=1,n} \min(\alpha_i, \mu_{B_i}(v))$$

i.e. here, with $a_{i-1} < a < a_i$

$$B = (\alpha_{i-1} \wedge B_{i-1}) \cup (\alpha_i \wedge B_i)$$

with $\mu_{\alpha \wedge B} = \min(\alpha, \mu_B)$. The conclusion B, obtained as a weighted union of the conclusions of the rules which more or less apply to the situation x = a (i.e. such that $\alpha_i = \mu_{A_i}(a) > 0$), is in agreement with the intended meaning of the rule. Indeed, the more x = a is in agreement with the condition part of rule number i, the more possible its conclusion ;

if several rules are applicable it yields several more or less possible (fuzzy) subsets for choosing the value of y. The union of these subsets provides a collection of values which are more or less possible for y. (Remember that in case of possibility qualification, the values outside the specified range are not excluded.)

5. Qualitative Reasoning Based on Fuzzy Arithmetics

Let us consider the two pieces of information "x is approximately equal to y" and "y is much larger than z". What can be said about z with respect to x ? This kind of situation can be for instance encountered in temporal reasoning when we know that two dates are close, while one of them takes place much after a third one. An approximate equality can be modelled by a fuzzy relation E of the form $\mu_E(u,v) = \mu_L(u - v)$, for instance

$$\forall u, \forall v, \mu_E(u,v) = \max\left(0, \min\left(1, \frac{\delta + \varepsilon - |u - v|}{\varepsilon}\right)\right) = \begin{cases} 1 & \text{if } |u - v| \leq \delta \\ 0 & \text{if } |u - v| > \delta + \varepsilon \\ (\delta + \varepsilon - |u - v|)/\varepsilon & \text{otherwise} \end{cases}$$

where δ and ε are respectively positive and strictly positive parameters which modulate the approximate equality. Similarly a more or less strong inequality can be modelled for instance by a relation I of the form

$$\forall v, \forall w, \mu_I(v,w) = \mu_K(v - w) = \max\left(0, \min\left(1, \frac{v - w - \lambda}{\rho}\right)\right) = \begin{cases} 1 & \text{if } v > w + \lambda + \rho \, ; \lambda \geq 0 \\ 0 & \text{if } v < w + \lambda \quad\quad ; \rho > 0 \\ (v - w - \lambda)/\rho & \text{otherwise} \end{cases}$$

The values of λ and ρ act on the meaning of the inequality and can express shades such as 'slightly greater', 'much greater', etc. When $\lambda = 0$ and $\rho \to 0$ the usual inequality relations '>' or '≥' are recovered (depending on how $\lim_{\rho \to 0} (v - w) / \rho$ is defined for $v \to w$). From $\pi_{x,y} = \mu_E$ and $\pi_{y,z} = \mu_I$, by combination and projection we get $\pi_{x,z} = \mu_{E \circ I}$, i.e.

$$\forall u, \forall w, \mu_{E \circ I}(u,w) = \sup_{v \in \mathbb{R}} \min(\mu_L(u - v), \mu_K(v - w))$$
$$= \sup_{\substack{s,t \\ u-w=s+t}} \min(\mu_L(s), \mu_K(t))$$
$$= \mu_{L \oplus K}(u - w)$$

where we recognize the expression of the extended sum of K and L (e.g. Dubois and Prade, 1988). Thus the possible values of the difference x − z is restricted by L ⊕ K. This result is represented in Figure 2 where the relations defined above are used. We see that it is certain that $x \geq z + \lambda − (\delta + \varepsilon)$ and that the value of the difference x − z belongs to L ⊕ K at the degree 1 as soon as $x \geq z + \lambda + \rho − \delta$. Then depending on the respective values of the parameters, x is still greater than z (but maybe not as such as y with respect to z) (if $\lambda − \delta − \varepsilon > 0$), or we are only sure that x is not much smaller than z (if $\lambda + \rho − \delta > 0$).

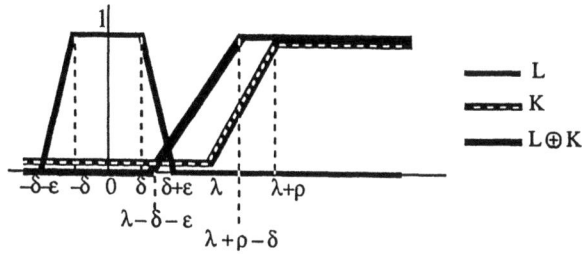

Figure 2

As it is suggested by this example, it is possible and even simple to derive inference rules in the case of fuzzy relations defined in terms of difference of values in order to perform these types of computation *at the symbolic level*. This can be done as well if we use the quotient for comparing numbers. For instance, let us suppose that "x is close to y" is understood as "x/y is close to 1", in the sense that the possible values of x/y are restricted by a fuzzy set F such that $\mu_F(1) = 1$, modelling "around 1". Similarly "z is negligible with respect to y" will be modelled by 1 + z/y is close to 1, where 'close' refers here to a fuzzy set G, possibly different from F, but still such that $\mu_G(1) = 1$. Then, in agreement with the combination/projection method, we can deduce at the symbolic level that, for instance, x + z is close to y in the sense of the fuzzy number F ⊕ G ⊖ 1 where ⊕ and ⊖ are the sum and the difference extended to fuzzy numbers (the extended difference is just defined by substituting − to + in the definition of ⊕). Indeed we have

$$\sup_{\substack{u,v \\ u+v=w}} \min(\mu_F(u), \mu_G(1 + v)) = \sup_{\substack{u,v \\ u+v=w}} \min(\mu_F(u), \mu_{G\ominus 1}(v)) = \mu_{F\oplus G\ominus 1}(w)$$

where $u = \frac{x}{y}$, $v = \frac{z}{y}$ and $w = \frac{x+z}{y}$. Thus the symbolic computation of the fuzzy relations relating quantities of interest can be performed independently of the practical semantical interpretation of these relations. In the above example, given the intended meanings of F and G, it is easy to compute what $F \oplus G \ominus 1$ means in terms of membership function. See Dubois and Prade (1991) for a detailed presentation of such a system of reasoning with relative orders of magnitude.

6. Multiple Criteria Aggregation with Twofold Fuzzy Sets

Apart from fuzzy rule-based inference and fuzzy arithmetics, an important application area of fuzzy sets is the modelling of objectives in decision processes and the use of the panoply of fuzzy set-aggregation operations in multiple criteria decision-making. This line of research was initiated by the pioneering work of Bellman and Zadeh (1970). A lot of works have been devoted to the study of fuzzy set operations and to their intended semantics (e.g. logical 'and' vs. compensatory 'and' ; see Zimmermann (1987)) ; see Dubois and Prade (1984, 1985) for reviews. Less efforts have been devoted by fuzzy set researchers to the way decision-makers build their decision especially in case of conflicting objectives, if we except Felix (1990, 1992). This author first distinguishes between alternatives which supports and alternatives which distracts a considered objective. Let $\mathcal{A} = \{a_1, ..., a_n\}$ be a set of potential alternatives, and $\mathcal{G} = \{g_1, ..., g_m\}$ be a set of objectives. To each objective g_i is associated a pair of fuzzy sets $(\overset{\circ}{G}_i, \overline{G}_i)$ where $\overset{\circ}{G}_i$ is the fuzzy set of alternatives which more or less certainly satisfy the objective g_i and the fuzzy set $\overline{G}_i$ of alternatives which more or less certainly do *not* satisfy this objective. Note that $\overset{\circ}{G}_i$ and $\overline{G}_i$ are not necessarily normalized. For the sake of consistency, the supports of $\overset{\circ}{G}_i$ and $\overline{G}_i$ are supposed to have no overlap (i.e. $\not\exists a$, $\mu\overset{\circ}{G}_i(a) > 0$ and $\mu\overline{G}_i(a) > 0$). The merit of this view is to distinguish between i) the alternatives preferred by the decision-maker with respect to objective g_i, ii) those more or less rejected, and iii) those in between (such that $\mu\overset{\circ}{G}_i(a) = 0 = \mu\overline{G}_i(a)$). Taking the complement G_i of $\overline{G}_i$ ($\mu G_i = 1 - \mu\overline{G}_i$), the pair $(\overset{\circ}{G}_i, G_i)$ is nothing but a twofold fuzzy set in the sense of Dubois and Prade (1987) ; G_i is the fuzzy set of alternatives which are more or less possible, tolerable according to the decision-maker.

Note that the support of $\overset{\circ}{G}_i$ is included in the core of G_i, and thus refines it (i.e. $\mu_{\overset{\circ}{G}_i}(a) > 0$ $\Rightarrow \mu_{G_i}(a) = 1$). The alternatives a such that $\mu_{G_i}(a) = 0$ can be considered as completely rejected by g_i, whatever the score of a with respect to other objectives ; in other words G_i expresses a *veto* condition with respect to criterion g_i (see Roy (1978) who stresses the importance of modelling vetos in decision processes). Thus G_i is a flexible constraint restricting the more or less tolerable alternatives according to g_i, while $\overset{\circ}{G}_i$ expresses the preferences of the decision-maker among the most satisfactory ones. Clearly conflicts between objectives are often encountered if we only consider the preferences (i.e. the $\overset{\circ}{G}_i$'s) ; it will be less often the case between the G_i's expressing the constraints.

Given two objectives g_i and g_j, several situations can be encountered (the following could be generalized to m objectives) :

i) $\overset{\circ}{G}_i \cap \overset{\circ}{G}_j \neq \emptyset$

ii) $(\overset{\circ}{G}_i \cap G_j) \cup (G_i \cap \overset{\circ}{G}_j) \neq \emptyset$ (but $\overset{\circ}{G}_i \cap \overset{\circ}{G}_j = \emptyset$)

iii) $G_i \cap G_j \neq \emptyset$ (but $\overset{\circ}{G}_i \cap G_j) \cup (G_i \cap \overset{\circ}{G}_j) = \emptyset$)

iv) $G_i \cap G_j = \emptyset$

where $\cap$ and $\cup$ denote fuzzy set intersection and union based on min and max respectively. In situation (i) there exists an alternative which is certainly satisfactory with respect to both objectives, while in situation (ii) the best alternatives are only preferable with respect to one of the objectives ; in situation (iii) it is only possible to find an alternative tolerable with respect to both objectives ; in the last situation any alternative is necessarily rejected to some degree by one of the objectives. A solution might then be to consider alternatives which are not too much rejected by one objective and which are more or less satisfactory according to the other. Felix (1990, 1992) has defined possible relationships between objectives like 'independence', 'cooperation', 'competition', on the basis of degrees of inclusion between $\overset{\circ}{G}_i$, G_i, $\overset{\circ}{G}_j$, G_j or their complements ; he proposes to apply heuristic rules like "if the objectives are independent or rather competitive then a disjunctive aggregation is recommended, while conjunctive aggregation is good for cooperative goals ", or " if g_i competes with g_j, sacrifice the less important objective. In case of conflicting goals (especially in situation (iv) above) another way is to introduce a priority ordering between the objectives according to their importance. It can be made by changing G_i into G^*_i with $\mu_{G^*_i} = \max(\mu_{G_i}, 1 - w_i)$ (where the weights are normalized, i.e. $\max_{i=1,m} w_i = 1$). In other words, alternatives out of the support of G_i are then considered as at least tolerable at the level $1 - w_i$; it corresponds to a kind of weighted veto (see Dubois and Koning, 1991).

For the sake of brevity, we have only considered max and min combination above. With the veto interpretation in mind, i.e. the understanding of G_i as a flexible constraint on feasible (tolerable) alternatives, a logical conjunction for the G_i's, i.e. with the min operation, seems natural (relaxing the G_i into G^*_i in case of severe conflicts); for the aggregation of the preferences, i.e. the $\overset{\circ}{G}_i$'s, a larger choice of operations may be considered ranging from logical 'and' to operations expressing trade-offs. However the aggregation process should preserve the requirement which holds for each objective, namely $\mu_{\overset{\circ}{G}_i}(a) > 0 \Rightarrow \mu_{G_i}(a) = 1$, for the result of the aggregation.

7. Conclusion

This paper has emphasized the capabilities of fuzzy set-based methods for modelling gradual properties, flexible constraints, preferences as well as for representing uncertain and imprecise pieces of information and various kinds of fuzzy rules. More particularly possibility theory offers a convenient framework for handling dual notions like certainty/possibility in case of incomplete information, or like satisfactory/tolerable in decision processes. The introduction of weights for assessing the certainty (and the guaranteed possibility) of statements, or the priority of flexible constraints modelling objectives has been also stressed. Finally, fuzzy set and possibility theory enables us to formalize a great variety of types of reasoning including propagation of uncertainty, interpolation, analogy, and qualitative computation in a deductive machinery based on the combination and the projection of possibility distributions.

References

Bellman R.E., Zadeh L.A. (1970) Decision-making in a fuzzy environment. Management Science, 17, B141-B164.

Berenji H.R. (1992) Fuzzy logic controllers. In : An Introduction to Fuzzy Logic Applications in Intelligent Systems (R.R. Yager, L.A. Zadeh, eds.), Kluwer Academic Publ., Boston, 69-96.

Dubois D., Koning J.L. (1991) Social choice axioms for fuzzy set aggregation. Fuzzy Sets and Systems, 43, 257-274.

Dubois D., Prade H. (1984) Criteria aggregation and ranking of alternatives in the framework of fuzzy set theory. In : Fuzzy Sets and Decision Analysis (H.J. Zimmermann, L.A. Zadeh, B.R. Gaines, eds.), Studies in the Management Sciences, Vol. 20, North-Holland, 209-240.

Dubois D., Prade H. (1985) A review of fuzzy set aggregation connectives. Information Sciences, 36, 85-121.

Dubois D., Prade H. (1987) Twofold fuzzy sets and rough sets – Some issues in knowledge representation. Fuzzy Sets and Systems, 23, 3-18.

Dubois D., Prade H. (with the collaboration of Farreny H., Martin-Clouaire R., Testemale C.) (1988) Possibility Theory – An Approach to Computerized Processing of Uncertainty. Plenum Press, New York.

Dubois D., Prade H. (1991) Semantic considerations on order of magnitude reasoning. In : Decision Support Systems and Qualitative Reasoning (M. Singh, L. Travé-Massuyès, eds.), North-Holland, Amsterdam.

Dubois D., Prade H. (1992a) Fuzzy rules in knowledge-based systems – Modelling gradedness, uncertainty and preference. In : An Introduction to Fuzzy Logic Applications in Intelligent Systems (R.R. Yager, L.A. Zadeh, eds.), Kluwer Academic Publ., Dordrecht, The Netherlands, 45-68.

Dubois D., Prade H. (1992b) Possibility theory as a basis for preference propagation in automated reasoning. Proc. of the 1st IEEE Inter. Conf. on Fuzzy Systems (FUZZ-IEEE'92), San Diego, CA, March 8-12, 821-832.

Felix R. (1990) Goal-oriented selection of alternatives based on fuzzy sets. Proc. of the extended abstracts of the 3rd Inter. Conf. on Information Processing and Management of Uncertainty in Knowledge-Based Systems (IPMU'90), Paris, July 2-6, 1990, 116-118.

Felix R. (1992) Towards a goal-oriented application of aggregation operators in fuzzy decision making. Proc. of the Inter. Conf. on Information Processing and Management of Uncertainty in Knowledge-Based Systems (IPMU'92), Mallorca, July 6-10, 585-588.

Klir G.J., Folger T.A. (1988) Fuzzy Sets, Uncertainty and Information. Prentice Hall, Englewood Cliffs, NJ.

Mamdani E.H. (1977) Application of fuzzy logic to approximate reasoning using linguistic systems. IEEE Trans. on Comput., 26, 1182-1191.

Roy B. (1978) ELECTRE III : un algorithme de classement fondé sur une représentation floue des préférences en présence de critères multiples. Cahiers du CERO, 20, 3-24.

Sugeno M., Nishida M. (1985) Fuzzy control of model car. Fuzzy Sets and Systems, 16, 103-113.

Yager R.R. (1983) An introduction to applications of possibility theory. Human Systems Management, 3, 246-269.

Zadeh L.A. (1965) Fuzzy sets. Information and Control, 8, 338-353.

Zadeh L.A. (1978) Fuzzy sets as a basis for a theory of possibility. Fuzzy Sets and Systems, 1, 3-28.

Zadeh L.A. (1979) A theory of approximate reasoning. In : Machine Intelligence, Vol. 9 (J.E. Hayes, D. Michie, L.I. Mikulich, eds.), Elsevier, Amsterdam, 149-194.

Zimmermann H.J. (1987) Fuzzy Sets, Decision Making, and Expert Systems. Kluwer Academic Publ., Boston.

The CALCULUS of Fuzzy If/Then RULES

BY LOTFI A. ZADEH

In contrast to classical logical systems, fuzzy logic is aimed at a formalization of modes of reasoning that are approximate rather than exact. Basically, a fuzzy logical system may be viewed as a result of fuzzifying a standard logical system. Thus, one may speak of fuzzy predicate logic, fuzzy modal logic, fuzzy default logic, fuzzy multivalued logic, fuzzy epistemic logic, and so on. In this perspective, fuzzy logic is essentially a union of fuzzified logical systems in which precise reasoning is viewed as a limiting case of approximate reasoning.

Recently, fuzzy logic has been finding a rapidly growing numer of applications in fields ranging from consumer electronics and photography to medical-diagnosis systems and securities-management funds. What is exploited in most of these applications is fuzzy logic's tolerance for imprecision. In effect, the operative principle of fuzzy logic is: precision is costly. Minimize the precision needed to perform a task.

As a simple illustration of this point, consider the problem of parking a car. Usually, a driver can park a car without too much difficulty, because the final position of the car is not specified exactly. If it were specified with high precision, it would take days and perhaps months to park the car. The same theory applies to HDTV. The NHK MUSE system and the recently proposed all-digital systems minimize the need for extended bandwidth by employing digital compression techniques that take advantage of fuzzy logic's tolerance for imprecision and knowledge of signal statistics.

Another case in point is the travelling-salesperson problem. As reported in *The New York Times* regarding recent advances in the approximate solution of this question, the time needed to solve the problem to within 1 % for 100,000 cities is two days; to within 0,75% for 100,000 cities is seven months; and to within 3,5 % for 1 million cities is three and a half hours. What is striking in this case is the dramatic effect of reducing the accuracy from 0,75 % to 3,5 %.

CALCULUS OF FUZZY RULES

Fuzzy logic provides a wide variety of concepts and techniques for representing and inferring from knowledge that is imprecise, uncertain, or unreliable. At this juncture, however, what is used in most practical applications is a relatively restricted and yet important part of fuzzy logic, centering on the use of fuzzy *if/then* rules. This part of fuzzy logic will be referred to as "the calculus of fuzzy *if/then* rules" (CFR) because it constitutes a fairly self-contained collection of concepts and methods for handling varieties of knowledge that can be represented in the form of a system of *if/then* rules in which the antecedents, consequents, or both are fuzzy rather than crisp. For example, expressed as a collection of fuzzy *if/then* rules, the relation between three variables, X, Y, and Z may be described as:

Z is medium if X is large and Y is not very small
Z is large if X is small and Y is medium
....
Z is medium if X is very small and Y is large

in which the linguistic values small, medium, and large are fuzzy sets. The meaning of such values is defined by their membership functions (Figure 1), which are specified by the designer or are induced from input-output data.

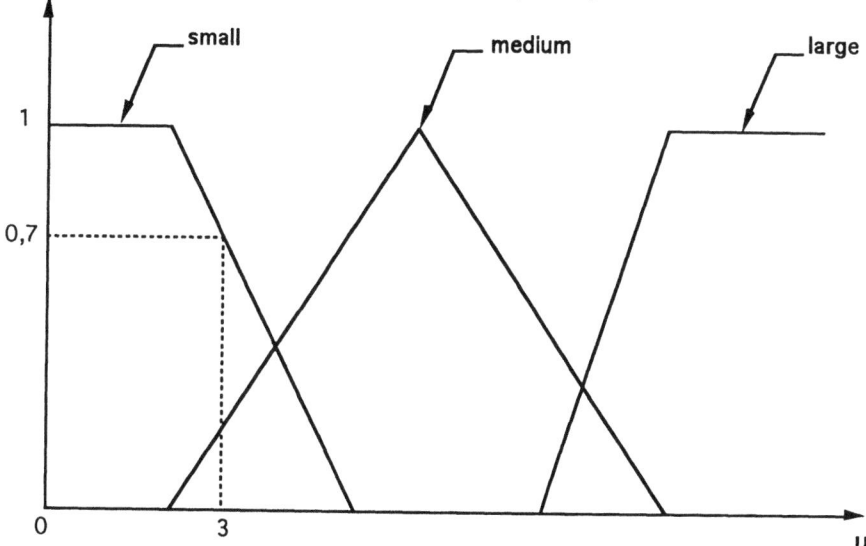

FIGURE 1.
Membership function of small medium, and large. 0,7 ist the grade of membership of 3 in small.

On a more concrete level, the relation between interest rate, unemployment and inflation may be expressed as a collection of fuzzy *if/then* rules exemplified by:

increase interest rates slighty if
unemployment is low and inflation is moderate

increase interest rates sharply if
unemployment is low and inflation is moderate but
rising sharply

decrease interest rates slightly if
unemployment is low but increasing and inflation
rate is low and stabile

The importance of the calculus of fuzzy *if/then* rules stems from the fact that much of human knowledge lends itself to representation in the form of a hierarchy of fuzzy *if/then* rules. Furthermore, the inference mechanisms in the calculus of fuzzy *if/then* rules are relatively simple and in harmony with the modes of human reasoning, which are approximate rather than exact. As a consequence, the calculus of fuzzy if/then rules is easy to master and apply.

What is of central importance in practical applications is that the fuzziness of the antecedents eliminates the need for a precise match with the input. As a consequence, in a fuzzy rule-based system, each rule is fired to a degree that is a function of the degree of match between its antecedent and the input. The mechanism of imprecise matching provides a basis for interpolation. Interpolation, in turn, serves to minimize the number of fuzzy *if/then* rules needed to describe an input-output relation. The bottom line is that fuzzy rule-based systems are simpler, cheaper, and more robust than their conventional counterparts.

The basic ideas underlying the calculus of fuzzy *if/then* rules were described in the late 1960s and early 1970s, and today, fuzzy *if/then* rules and their applications are covered extensively in available literature. However, the calculus of fuzzy *if/then* rules is much more than a summary of the literature. What it represents is, in effect, a crystallization of a self-contained branch of fuzzy logic in which the concept of a fuzzy *if/then* rule plays a central role.

In this perspective, the agenda of the calculus of fuzzy *if/then* rules may be set down as follows:

- Interpretation of a fuzzy *if/then* rule
- Interpretation of a collection of fuzzy *if/then* rules
- Representation of propositions in a natural language as collections of fuzzy *if/then* rules
- Inference from a collection of fuzzy *if/then* rules (interpolation)
- Manipulaton of blocks of fuzzy *if/then* rules
- Algebraic operations on fuzzy *if/then* rules
- Induction of fuzzy *if/then* rules from observations
- Applications
- Control
- Qualitative systems analysis
- Qualitative decision analysis
- Signal processing and data compression

Our aim in this article is to outline some of the basic ideas underlying the calculus of fuzzy if/then rules and point a way to their applications.

FUZZY IF/THEN RULES

Typically, a fuzzy *if/then* rule has this format:

Y_1 is B_1 and Y_2 is B_2 ... and Y_m is B_m if
X_1 is A_1 and X_2 is A_2 ... and X_n is A_n

where the As and Bs are linguistic values, such as *small, large,* and *not very small.* The left-hand side of the rule is the consequent and the right-hand side is the antecedent. In the simplest case, n = m = 1. For our purposes, it will be sufficient to focus on this case, so that a rule may be expressed as:

Y is B if X is A.

More generally, a rule may be qualified:

usually *(Y is B if X is A)*

or:

usually (Y is B) if X is A

or:

Y is B if X is A unless X is E

Furthermore, a rule may have exceptions, as in:

Y is B if Y is A unless X is E

Such rules will not be condidered in this article.

The point of departure in the calculus of fuzzy if/then rules is the interpretation of a single rule. Consider the rule:

Y is B if X is A

in which X and Y are variables whose domains are U and V, respectively, and A and B are fuzzy predicates or relations in U and V, which play the role of elastic constraints on X and Y. For example:

Volume is low if Pressure is high

In this case, V = Volume, X = Pressure, B = low, A = high.

In fuzzy logic, a fuzzy if/then rule may be interpreted in several ways. In most applications, the rule "Y is B if X is A" is interpreted as a specification of the joint-possibility distribution of X and Y, that is:

$$\Pi(X, Y) = A \times B$$

where $A \times B$ represents the cartesian product of A and B. This specification implies that the possibility-distribution function of X and Y may be expressed as:

$$\pi(u, v) = \mu_A(u) \; \lambda \; (v) \; \mu_B(v)$$

where μ_A und μ_B are the membership functions of A und B, respectively, and λ denotes the min operator. Given the rule, the possibility that $X = u$ and $Y = v$ is given by $\mu_A(u) \lambda(v)\mu_B(v)$. A graphical representation of this interpretation is shown in Figure 2.

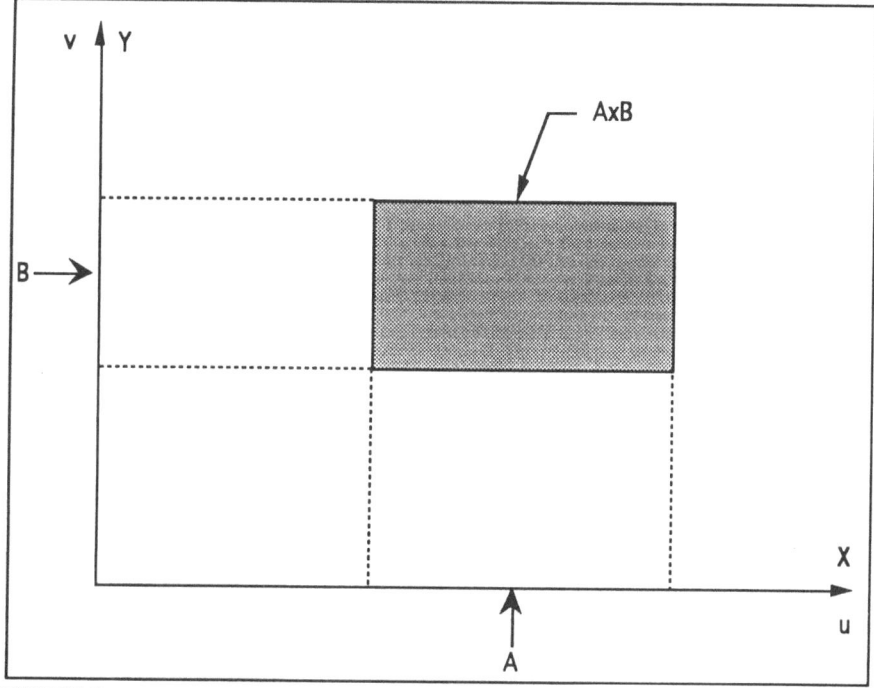

FIGURE 2.
The joint possibility distribution of X and Y defined by the rule "Y is B if X is A".

A basic item on the agenda is: How should two or more rules be combined? In the case of the joint-possibility distribution, the interpretations are combined disjunctively. More specifically, a collection of fuzzy *if/then* rules of the form:

· Y is B_i if X is A_i, $i = 1,...,n$

defines the joint-possibility distribution:

$$\Pi(X,Y) = A_1 \times B_1 + ... + A_n \times B_n$$

in which + plays the rule of disjunction. Thus, if the cartesian product $A_i \times B_i$ is visualized as a fuzzy point, $\Pi(X, Y)$ may be viewed as a fuzzy graph or fuzzy relation R, which can be expressed as

$$R = \sum_i (A_i \times B_i)$$

A graphical representation of this intepretation is shown in Figure 3.

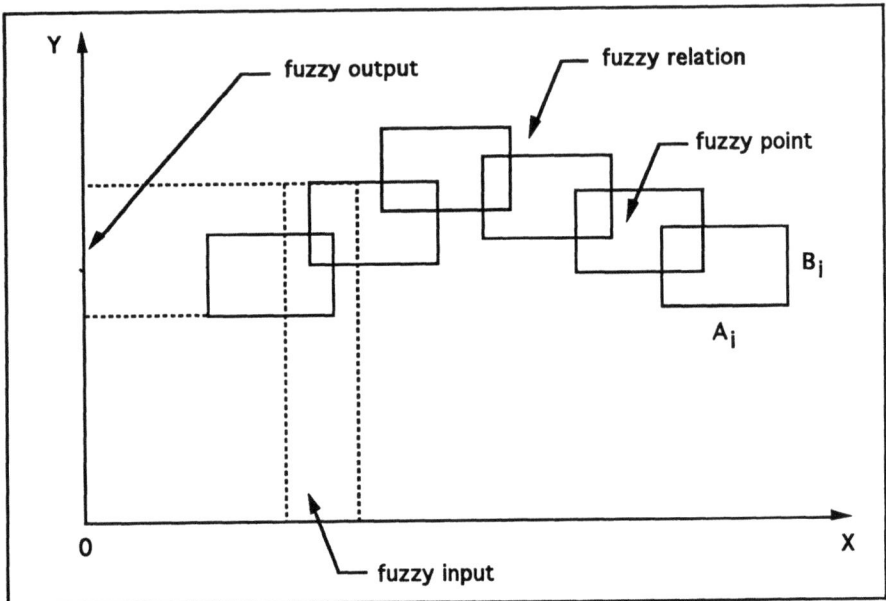

FIGURE 3
The fuzzy graph of a collection of fuzzy points each of which represents a fuzzy *if/then* rule.

In the notation of FA-Prolog, the relation in question may be written as:

$r(X, Y; \lambda) : - a_1(X; \mu_1), b_1(Y; \nu_1); ...; a_n(X; \mu_n), b_n(Y; \nu_n)$

in which λ is the truth value of the fuzzy predicate r and $\mu_1,...,\mu_n, \nu_1,...,\nu_n$ are the truth values of the predicates $a_1,...,a_n, b_1,...,b_n$, respectively. (The lower-case predicate a_i is a Prolog representation of A_1.)
For purposes of interpolation, given an input "X is A", the corresponding output, "Y is B", may be expressed in FA-Prolog as:

$b(Y; \lambda) : - r(X, Y; \mu), a(X; \nu).$

COMBINATION OF BLOCKS OF RULES

In qualitative systems analysis and intelligent control, a fuzzy rule set may be interpreted as a qualitative description of the input-output relation of a system. Thus, a rule set of the form:

Y is B_1 if X is A_1

Y is B_2 if X is A_2

....

Y is B_n if X is A_n

may represent the input-output relation of a system R, with the relation in question expressed as the fuzzy relation:

$$R(X, Y) = \sum_i (A_i \times B_i).$$

In the case of a serial combination of $R1$ and $R2$, the problem is to derive the input-output relation of the serial combination, $R12$, from the knowledge of the input-output relations $R1$ and $R2$. In FA-Prolog, we now have:

$r1(X, Y; \lambda) : -a_1^1(X; \lambda_1), b_1^1(Y; v_1); ....; a_n^1(X; \lambda_n), b_n^1(Y; v_n).$

$r1(Y, Z; \lambda) : -a_1^2(Y, \lambda_1), b_1^2(Z; v_1); ...; a_n^2(Y; \lambda_n), b_n^2(Z; v_n).$

The solution to the problem can be expressed as:

$r12(X, Z; \lambda) : - r1(X, Y; \mu), r2(Y, Z; v).$

In a more general situation, involving the presence of feedback loops, the input-output relation may be expressed as:

$r12(X_1, Y_1; \lambda) : - r1(X_1, X_2, Y_1, Y_2; \mu), r2(Y_2, X_2, v).$

In the case of $R1$, the fuzzy if/then rules are of the form:

Y_1 is B_1^1 and Y_2 is B_1^2 if X_1 is A_1^1 and X_2 is A_1^2

and the corresponding fuzzy input-output relation may be expressed as:

$$R1\,(X_1, X_2, Y_1, Y_2) = \sum_i A_i^1 \times A_i^2 \times B_i^1 \times B_i^2.$$

The same approach can be used to compute the input-output relation of any combination of systems, each of which is characterized by a collection of fuzzy *if/then* rules.

It can be used to analyze, in qualitative terms, the behavior of a complex system whose constituents are characterized by fuzzy rule sets with linguistic variables rather than algebraic or differential equations. Important application areas for this approach are industrial-process control, prediction and signal processing, and large-scale system modeling.

Another important application area relates to probabilistic systems in which the underlying probabilities are not known with sufficient precision to be treated as numbers. The calculus of fuzzy *if/then* rules provides a basis for handling such probabilities as collections of fuzzy *if/then* rules. More specifically, assume that X and Y are random variables whose probability distributions are described in linguistic terms. For example:

$p(X)$: *probability is low if X is small*
probability is high if X is medium
probability is low if X is large
....
$q(Y|X)$: *probability is high if X is small and Y is large*

where $q(Y|X)$ is the conditional probability distribution of Y, given X. The problem is to compute the probability distribution of Y in the form of a collection of fuzzy *if/then* rules with linguistic variables.

Since computation with probabilities involves the operations of addition and multiplication, it is necessary to have a method for operating on functions characterized by collections of fuzzy *if/then* rules. For example, if $f(X)$ and $g(Y)$ are described in terms of such rules as disjunctions of fuzzy points, then $f(X)$ and $g(Y)$ may be expressed as:

$$f(X) = \sum_i (F_i \times F_i')$$

and

$$g(Y) = \sum_i (G_j \times G'_j).$$

Furthermore, if $f(X)$ and $g(Y)$ are combined through a binary operation such as multiplication or addition, the resulting function may be expressed as:

$$h(X, Y) = f(X) * g(Y).$$

The fuzzy rule set associated with $h(X, Y)$ is given by

$$h(X, Y) = \sum_{i,j} (F_i \times G_j) \times (F'_i * G'_j)$$

This result provides the basis for computing imprecisely-known probabilities expressed as collections of fuzzy *if/then* rules involving linguistic variables. In the case of such probabilities, we have:

$$p(X) = \sum_i (P_i \times P'_i)$$

$$q(Y|X) = \sum_j (Q_j \times Q'_j)$$

$$r(X, Y) = p(X) q(Y|X)$$

In the application of the calculus of fuzzy *if/then* rules, one problem is the calibration of the membership functions of the linguistic values. The practice in the design of fuzzy-logic-based systems employing such rules involves, for the most part, the use of cut-and-trial procedures. Recently, however, steps have been taken toward the development of adjustment algorithms - some of which are based on neural network techniques - for a systematic approach to the problem of calibration.

In summary, the calculus of fuzzy *if/then* rules provides a systematic way of handling systems in which it is either necessary or advantageous to describe the input-output relations in the form of fuzzy *if/then* rules.

The calculus of fuzzy *if/then* rules is simple and close to intuition. Furthermore, it is largely self-contained and does not require an extensive familiarity with fuzzy logic. For these reasons, it is likely to become a widely used tool in systems analysis, control, signal processing, pattern recognition, decision analysis, diagnostics, and related fields.

SUGGESTED READING

Dubois, D. and H. Prade
Possibility theory: An approach to computerized processing of uncertainty. New York: Plenum Press, 1988

Pedrycz, W.
Fuzzy control and fuzzy systems. Somerset, England: Research Studies Press Ltd., 1989

Zadeh, L. A.
"Fuzzy algorithms", Inf. Control 12, 94-102, 1968

Zadeh, L. A.
"Toward a theory of fuzzy systems", Aspects of Network and System Theory, R. E. Kalman and N. DeClaris (eds.), New York, N.Y.: Rinehart and Winston, 1971
pp. 469-490,

Zadeh, L. A.
"Outline of a new approach to the analysis of complex systems and decision processes", IEEE Trans. on Systems, Man and Cybernetics, SMC-3, 28-44, 1973

Zadeh, L. A.
"On the analysis of large scale systems," Systems Approaches and Environment Problems. H. Gottinger (ed.) Göttingen, Germany: Vandenhoeck and Ruprecht, pp. 23-37, 1974

Zadeh, L. A.
"QSA/FL-Qualitative systems analysis based on fuzzy logic", Stanford, Calif.: Stanford AAAI Symposium on Limited Rationality, pp. 111-114, 1989

Implementation of Fuzzy Technology

V.A.M. Kwaks and B.W. Grant

OMRON - European Technical Centre
Zilverenberg 2
5234 GM 's-Hertogenbosch
The Netherlands
Tel: +31 73 481811

ABSTRACT

In this paper implementations of control systems using fuzzy logic are discussed. The different implementation methods are compared and strengths and drawbacks are determined and discussed.

Introduction

During the last few years a growing interest in fuzzy logic in Europe has been seen. The variety of Japanese consumer products utilising this technology[1],[2] and the reports on successful applications of fuzzy technology (especially in Japan) have stimulated the interest. Applications can be found in various fields ranging from industrial process control[3] to medical diagnosis[4] and securities trading.

Fuzzy logic systems base their decisions on inputs in the form of linguistic variables represented by so-called membership functions. These functions are a mathematical representation of the associated linguistic value in the input interval. Using these linguistic variables IF..THEN rules can be formulated expressing expert knowledge and enabling fuzzy reasoning[5],[6].

To perform this fuzzy reasoning several fuzzy operators have to be defined. In control applications the minimum (MIN) and maximum (MAX) operators are mostly used. Fuzzy inference results in a fuzzy conclusion which has to be defuzzified. The centre of gravity method (normalised integral over output interval) is most often used in control applications.

The implementation of fuzzy logic using standard processors is not the most elegant and certainly not the most efficient way. However one can imagine that the besides technical issues, economical issues (consumer products, automotive applications) play an important role.

Implementation of Fuzzy Technology: the possibilities

The following figure illustrates some of the options which are currently available for implementing fuzzy logic. Each of these options have corresponding advantages and disadvantages. The final implementation choice will be made by weighing up these advantages and disadvantages within the requirements of the control process. Economical considerations also have an effect on the choice of implementation in most applications and the final design decision will be a trade-off between all of these considerations.

In this section we will discuss the physical implementation details and in the next section the effect of the economical considerations on the final implementation choice will be discussed.

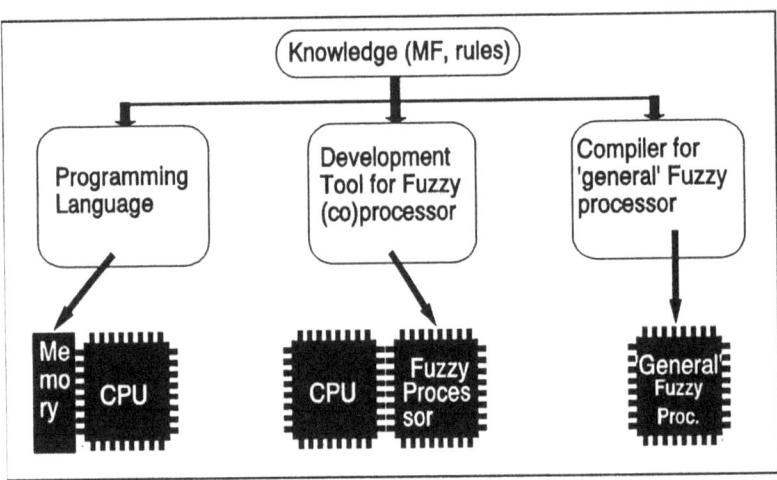

Figure 1 Implementation of Fuzzy Technology

a) Software Implementation

Fuzzy sets and fuzzy operations, such as fuzzy MIN, fuzzy MAX, fuzzy bounded SUM etc., are mathematically well defined[5],[6]. This allows fuzzy systems to be modelled. The easiest way to model a fuzzy system and one which requires no specialised hardware, is to implement it in software.

In a software implementation the model is written in a high-level computer language and compiled down to the machine language of a standard microprocessor using existing compiler software. All aspects of a fuzzy controller(fuzzification, inference and defuzzification) can be modelled in software. The advantages offered by a software implementation of a fuzzy system are flexibility and ease of use.

Software is highly configurative. It can be quickly and easily changed at relatively no cost at all. This is important in the development of fuzzy controllers when it is often not known in advance how many inputs are critical to the process and how many rules and membership functions will be needed to correctly model the process. The flexibility of the software approach allows the model to be changed when necessary without the expense of re-designing and re-implementing physical components of the system.

The cost of this flexibility is inferencing speed. The host CPU has to devote a proportion of it's processing power to take care of system requirements. The host CPU will not have specially defined fuzzy operations as part of it's instruction set, these operations must be defined by multiple calls to available CPU instructions. This results in a processing overhead which is high and a corresponding processing speed which is low.

In many applications the required fuzzy inference speed is low and it may be possible that a software implementation might be sufficiently quick to model your fuzzy system, however, in the case of many embedded control applications, where the inference speed required is high, a software model can prove to be inadequate. In these applications a specialised hardware implementation is required.

b) Dedicated Fuzzy Circuitry

By far the most common approach to implementing fuzzy operations in hardware is that of the ASIC fuzzy co-processor.

The relatively slow inferencing capability of software can be solved in hardware. By using specially designed hardware units to implement fuzzy operations it is possible to process fuzzy inputs in parallel. This dramatically improves the inference speed of the fuzzy model, therefore it comes as no surprise that a considerable amount of research effort is devoted to exploring the possibilities of implementing fuzzy sets and operations in hardware.

The fruits of this research activity has been the development of a number of commercially available fuzzy microprocessors from, among others, Omron, Togai and Neuralogix. These fuzzy co-processors have been successfully applied to a variety of control applications from small-scale consumer products, such as video camera stability controllers, to larger industrial applications, such as temperature controllers and industrial robots.

Application specific fuzzy co-processors can be broadly divided into two categories: analogue systems, as studied by Ueno[8] and Yamakawa[9] and digital systems, as studied by Watanabe[11] and Corder[7].

In analogue systems fuzzy operations MIN and MAX can be modelled from simple transistor circuits. Once the number of inputs and outputs have been decided upon a circuit can be constructed to perform fuzzy MIN-MAX composition from these basic circuit units. These circuits perform extremely fast parallel inferences. They have no requirement to convert analogue input signals to digital before processing, to store results in memory during processing, or to convert digital outputs to analogue control signals after processing, thus making them extremely efficient[8].

Among the most notable digital techniques currently used for implementing fuzzy operations in hardware are: Fuzzy Programmable Logic Arrays[9], Programmable Gate Arrays[10] and CMOS VLSI circuitry[11].

As in the analogue hardware, these solutions are based on creating specialised circuitry to perform specific fuzzy operations. With PLA's and PGA's the hardware can be created by hardware development "compilers" which create the required chip layout given a high-level specification.

Application specific fuzzy functional units, whilst providing specialised hardware to perform fuzzy operations, do not have the necessary circuitry to enable them to function as stand-alone units. They must be used in conjunction with standard micro-controllers or as co-processors for general microprocessors. This greatly speeds up the control process as now the microprocessor is free to perform I/O and memory management tasks while leaving the processor-intensive fuzzy operations to the specially designed fuzzy co-processors.

An important consideration which must be taken into account when using co-processors is that of interface capability. To be useful, and marketable, they must be able to interface directly with a wide variety of standard CPU's. This is easily done by providing access to the fuzzy units through memory. They knowledge base is loaded into the system memory at a pre-determined address. The CPU sends a signal to the fuzzy co-processor telling it that the knowledge data is available and to start inference. The fuzzy co-processor performs inference on the knowledge base while the CPU is free to perform other tasks. When inference is complete the fuzzy co-processor writes the results to memory and sends an interrupt to the host CPU to indicate that the result is available. A control action can now be performed based on the result.

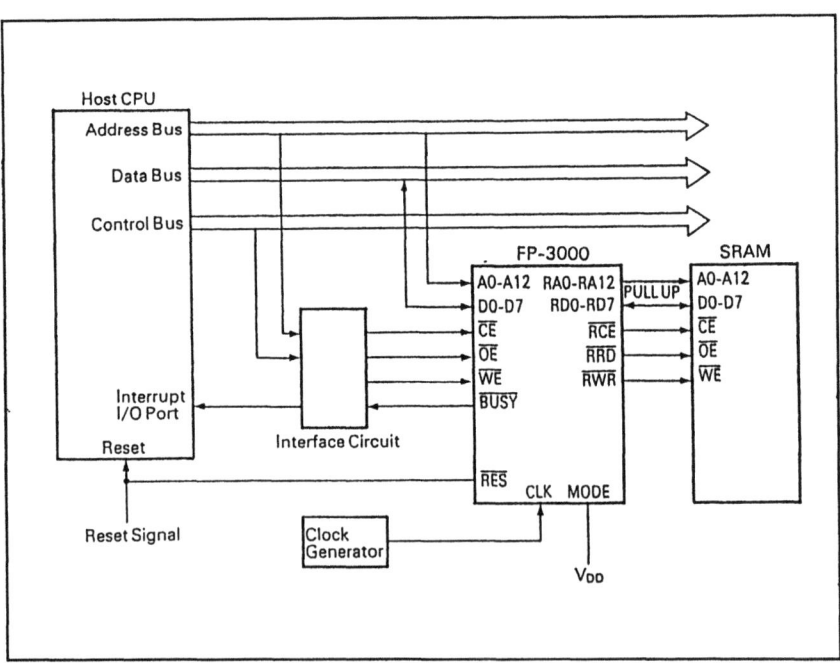

Figure 2. OMRON's FP-3000 configuration (Expanded mode)

Fuzzy co-processing units provide an easy upgrade path for existing conventional control systems. Often the fuzzy unit can be added to an existing microcontroller without too much effort, allowing existing systems to access the benefits of fuzzy control.

The disadvantage of these custom-designed circuits is that they have fixed architectures. The resolution of the inputs/outputs (if digital), the defuzzification method and the inference method are all fixed during the design process.

c) Fuzzy Processors

Processors incorporating fuzzy functional units with special instructions sets, to access these fuzzy functional units, have been developed. These fuzzy processors also have the functionality of standard processors with the ability to communicate with the other peripherals and to manage it's own memory units. This frees these processors from the constraints of interfacing with host processors therefore allowing them to be used as stand-alone units.

These stand-alone fuzzy control processors are commonly implemented as Reduced Instruction Set Computers (RISC's)[12]. Although they differ from the conventional RISC paradigm of providing a small number of basic instructions as building blocks for more complex operations. Instead, these RISC's provide a small number of complex, highly specific instructions. They are able to do this because the application area in which these processors are used is well-defined. This means that instructions usually found in CISC (Complex Instruction Set Computer) architectures which are needed to perform more general tasks not usually associated with control can be omitted. This results in a highly tailored instruction set containing only instructions for performing fuzzy operations and for communicating with external devices.

d)   Example of a product range: OMRON's Fuzzy Co-processors

The following figure illustrates the range of fuzzy co-processors developed by OMRON. One must realise that only a few of these processors are available or will be available on the European market.

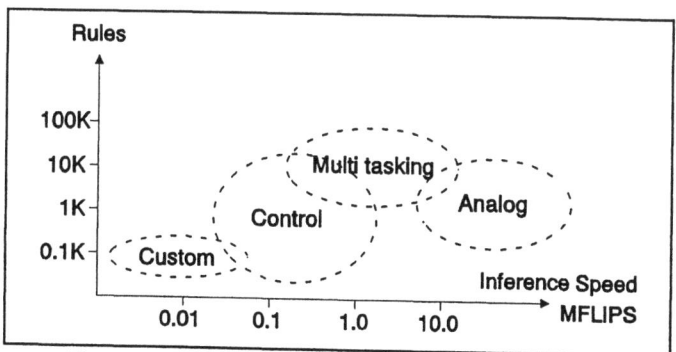

Figure 3. OMRON's range of fuzzy co-processors

Every application area imposes different requirements[13] on the fuzzy co-processor directly reflected in the figure above. The customised fuzzy processor (FP-1000) will be mainly used to assist in small and simple control systems using few inputs and generating few outputs. The number of rules is limited and the execution speed of the inference is not critical for these applications. Examples are washing machine[1] (consumer products), anti-skid braking systems and transmission control systems[13] (automotive).

One of the areas where fuzzy logic is an alternative for solving existing problems is control. A fuzzy co-processor (FP-3000[14], marketed in Europe) for these applications has been developed. It offers a 12-bit resolution and a simple interfacing towards the main CPU. Defuzzification method is user-selectable and 4 different shapes can be used to create membership functions. The wide range of possible applications in the field of control imposes different requirements on these co-processors requiring some flexibility e.g. selectable defuzzification method. The number of rules however will be limited. The knowledge base for these systems is in most applications a representation of operator knowledge. It is clear that the number of rules used by these operators is not infinite. The speed of the inference can become an important issue especially for real-time control (e.g. robot controller, conveyer control).

If more (fuzzy) processing power and knowledge storage capacity is required other problems arise. Co-processors which offers this functionality have been developed, but one of the problems still existing is the determination of the required knowledge for these systems. These processors are able to handle up to about 32000 rules. Tools for the automatic creation of rules are required before efficient usage of these co-processor becomes possible. Combining the learning abilities of neural networks to the development and tuning of fuzzy logic based systems is one major area of research[15].

The last in the processor range is the recently developed analogue processor. The proposed usage of these processors can be regarded as similar to that of transputers. These analogue processors can achieve enormous speed, but the rule storage capacity is limited. Parallelism is certainly required here introducing further complexity.

Implementation of Fuzzy Technology: the right choice

In this section economical aspects are considered. These directly influence the choice of implementation of fuzzy technology as described previously.

The software option requires a huge amount of computing power. Although fast-CPU and memory are becoming less expensive, the cost for a minimum required configuration can certainly not be neglected. Also the software required, which produces compilable code for the fuzzy controller can be quite expensive. Possible problems arise in some application areas. In control Programmable Logic Controllers (PLC) are widely used. The main CPU in these systems is not suitable for execution of the software calculating the fuzzy inference. A special module with a dedicated processor is needed. One has of course the possibility to use a standard processor in combination with the created source code for the fuzzy controller, but a more efficient choice would be to use a fuzzy co-processor. To enable communications in these systems between the main-CPU and the connected units, processors executing a communicating protocol are used. This processor can normally also be used to communicate with the fuzzy co-processor. An example of such a unit is OMRON's C200H-FZ001, the Fuzzy unit for the C200H PLC range using the FP-3000 fuzzy co-processor.

This illustrates one of the strengths of the second option. Fuzzy co-processors can offer design flexibility if they are available over a wide range of functionality and cost. The main advantage is that these chips can be added to existing processor controlled systems to achieve higher functionality or create a superior product without redesign. The only changes consist of creating a interface to the fuzzy co-processor.

The final option has similar advantages. If the instruction set of the standard processor is not compromised and pin-compatibility remains, then only adjustments are required to the software to incorporate fuzzy operations in the controller. However some drawbacks may exist, e.g. reduced performance introduced by the overhead to execute the fuzzy operations. The incorporation of the fuzzy instructions in the instruction set causes a trade off between existing functionality, incorporation of fuzzy technology and cost.

Conclusion

The implementation of fuzzy technology in Europe is still at its infancy. The wide range of possible applications make it difficult to clearly define a standard method to implement fuzzy technology. In this article several possible implementation methods have been discussed. Which implementation method to use in a certain application is dependent on both technical and economical constraints. The decision process can be broadly divided into two categories.

- The first starts from the current situation and more importantly systems currently being used. If enough processing power is available and speed is not critical then the software option is the best to start off with. If however there is an existing system working at its peak performance then a fuzzy co-processor might be the best solution. Depending on the required functionality the general fuzzy processor option is also a good alternative.

- The other situation is where there is no existing system. One should then define all the requirements (technical but also economical) and use these as the selection criteria for the implementation choice. The advantages and disadvantages of the possible implementations described in this paper may function as a guideline.

References

[1] KONDOH, S., ABE, S., TERAI, H., KIUCHI, M. and IMAHASHI, H.
"Fuzzy Logic Controlled Washing Machine"
Proceedings of International Workshop on Fuzzy System Applications (IFSA), Vol. Engineering, pp. 97 - 100,
1991, Brussels, Belgium.

[2] EGUSA, Y., AKAHORI, H., MORIMURA, A. and WAKAMI, N.
"An electronic video camera image stabilizer operated on fuzzy theory"
Proceedings of the IEEE International Conference on Fuzzy Systems 1992, pp. 851 - 858,
8 - 12 March 1992, San Diego, USA.

[3] SAITO, Y. and ISHIDA, T.
"Fuzzy PID Hybrid Control, An application to burner control"
Proceedings of the International Conference on Fuzzy Logic & Neural Networks, Vol. 1, pp. 65 - 69
20 - 24 July 1990, Iizuka, Japan.

[4] MEIER, R., NIEUWLAND, J., HACISALIHZADE, S., STECK, D. and Zbinden, A.
"Fuzzy Control of blood pressure during anestesia with Isoflurane"
Proceedings of the IEEE International Conference on Fuzzy Systems 1992, pp. 981 - 988,
8 - 12 March 1992, San Diego, USA.

[5] DUBOIS, Didier and PRADE, Henri
"Fuzzy Sets and Systems, Theory and Applications"
Mathematics in Science and Engineering, Vol. 144
Academic Press, 1980.

[6] ZIMMERMAN, H.J.
"Fuzzy Set theory and it's applications, second edition"
Kluwer Academic Publishers, 1991.

[7] CORDER, R.
"A High Speed Fuzzy Processor"
Proceedings of the 3rd IFSA Conference, pp. 779 - 381,
6 - 11 August 1989, Seattle, USA.

[8] UENO, F., INOUE, T., MAMORU, S., MORIMOTO, T.
"A Fuzzy Inference Engine Using Current-mode Max-Min Composition Circuits"
Proceedings of the 3rd IFSA Conference, pp. 639 - 642,
6 - 11 August 1989, Seattle, USA.

[9] YAMAKAWA, Takeshi
"A Fuzzy Programmable Logic Array"
Proceedings of the IEEE International Conference on Fuzzy Systems 1992,
pp. 459 - 465, 8 - 12 March 1992, San Diego, USA.

[10] MANZOUL, M.A. and JAYABHARATHI, D.
"Fuzzy Controller on a FPGA Chip"
Proceedings of the IEEE International Conference on Fuzzy Systems 1992,
pp. 1309 - 1316,
8 - 12 March 1992, San Diego, USA.

[11] WATANABE, H., DETTLOF, W.A. and YOUNT, K.E.
"VLSI Chip for Fuzzy Logic Inference"
Proceedings of the IFSA Congress, pp. 292 - 295,
6 - 11 August 1989, Seattle, USA.

[12] WATANABE, H.
"RISC Approach to Design of Fuzzy Processor Architecture"
Proceedings of the IEEE International Conference on Fuzzy Systems 1992,
pp. 431 - 440,
8 - 12 March 1992, San Diego, USA.

[13] IKEDA, H., KISU, N., HIRAMOTO, Y. and NAKAMURA, S.
"A Fuzzy Inference Coprocessor using a flexible active-rule-driven architecture"
Proceeding of the IEEE International Conference on Fuzzy Systems 1992,
pp. 537 - 544,
8 - 12 March 1992, San Diego, USA.

[14] KINOSHITA, I., KITA, S. and EJIMA, H.
"Digital Fuzzy Processor FP-3000" (In Japanese)
Proceedings of the 7th Fuzzy System Symposium, pp. 149 - 152,
12 - 14 June 1991, Nagoya, Japan.

[15] LIN, C. and LEE, G.
"Neural-Network-Based Fuzzy Logic control and decision system"
IEEE Transactions on Computers, Vol 40, No. 12
December 1991, pp. 1320 - 1336.

Teil 2

MITSUBISHI Fuzzy-Logic im Werkzeug- und Formenbau

Praxiseinsatz in der Funkenerosionstechnik

Betr.-Wirt. (VWA) Otto Keilhofer
MITSUBISHI ELECTRIC EUROPE GmbH,
Ratingen

1. Die Funkenerosion im Werkzeug- und Formenbau

1.1. Einsatzgebiete

Überwiegendes Einsatzgebiet ist die Bearbeitung komplexer und schwieriger, unter Umständen nicht mehr spanend zu fertigender Materialien. Es handelt sich dabei meist um eine Endbearbeitung bei der es auf höchste Präzision und äußerste Sorgfalt ankommt. Die Oberflächenstruktur, die sich aus der funkenerosiven Bearbeitung ergibt, kann z.B. die Struktur ("Orangenhaut") auf dem Kunststoffteil sein, das mit diesem Formwerkzeug gespritzt wird.

1.2. nötige Hilfsmittel

Die gebräuchlichsten Elektrodenmaterialien sind Kupfer, Graphit und deren Legierungen. Für die Arbeitselektroden und deren Fertigung Elektrodenhalter und -aufnahmen, Werkzeugmaschinen wie Fräs-, Dreh-, Schleifmaschinen.

1.3. Wirtschaftlichkeit gegenüber anderen Fertigungsverfahren

In vielen Fällen sind keine teueren Werkzeuge nötig; durch Einsatz von einfachen Kupferhalbzeugen, in Verbindung gebracht mit den Möglichkeiten einer offenen 4-Achsen-CNC- Steuerung, können teuere, spanabhebende Werkzeuge (Spezialfräser) ersetzt werden.

Ein hervorzuhebendes Einsatzgebiet ist dabei die Herstellung komplizierter Formwerkzeuge mit in Form gebogenen Konturen für Rohrprofile oder einfache Kupferbleche für Dichtungsprofile. Oftmals sind bei bestimmten Konturen polierte Endoberflächen nötig. Sie sind durch Standardeinstellungen aus einem umfangreichen Technologiekatalog wirtschaftlich zu erzielen.

2. Probleme in der Funkenerosion

2.1. Beschreibung des Verfahrens

2.1.1. Senkerosion

Eine vorhandene Formelektrode wird durch einfahren (einsenken) in achsialer Richtung (X, Y,Z) in das zu bearbeitende Werkstück übertragen. Ein rein abbildendes Verfahren, positiv/negativ.

2.1.1. Schneiderosion

Mittels eines dünnen Drahtes (ab 0.03 mm bis 0.3 mm Durchmesser) und einer 2- bis 5-Achsen-Steuerung wird die beschriebene Kontur (NC- Programm) in das Werkzeugmaterial geschnitten.

2.1.3. Funkenerosives Schleifen

Eine vorhandene Form- oder Halbzeugelektrode wird durch einfahren und (gleichzeitiger) Planetarbewegung in das zu bearbeitende Werkstück übertragen. Neben dem abbildenden Verfahren durch Einsatz der 3-/4-Achsen CNC-Steuerung herrscht eine ständige Bewegung (mit der Arbeitselektrode), dadurch entsteht ein gewünschter "schleifender" Effekt. Diese Arbeitsweise ist entsprechend wirtschaftlich, weil ein auftretender Elektrodenverschleiß auf die gesamte Eingriffsfläche der Elektrode verteilt wird und durch das günstigere Verschleißverhalten häufig mit einer geringeren Anzahl von Elektroden auch Mehrfachwerkzeuge zu fertigen sind.

2.2. Spezielle Problematik beim Senkerodieren

An zwei elektrisch leitende Körper (Elektrode und Werkstück) wird eine elektrische Spannung angelegt. Der Zwischenraum zwischen diesen beiden Elektroden wird mit einer schwach leitenden Flüssigkeit (Dielektrikum) aufgefüllt. Wenn die beiden Elektroden zueinander gefahren werden, kommt es bei einem bestimmten Abstand (Funkenspalt) zu einem Funkenüberschlag. Der Strom beginnt zu fließen.

Das Dielektrikum kühlt, reinigt und hält eine gewisse Isolation aufrecht, damit nur gezielte, bewußt gewollte Kurzschlüsse (Funkenüberschläge) auftreten. Damit ist eine hohe Maßhaltigkeit im Endergebnis zu erreichen.

2.3. Lösungswege

2.3.1. Adaptive Kontrolle

Die adaptive Kontrolle tritt dann ein, wenn bestimmte Vorgänge unkontrolliert ablaufen und eine Selbstregelung nicht mehr funktioniert. Sie läuft nach einem Schema ab, das in der Kontrolleinrichtung der modernen Senkerodieranlage festgeschrieben ist.

2.3.2. Automatik-Regelung

Die automatische Regelung vollzieht sich digital (EIN/AUS), was, in welcher Situation, wie zu regeln ist und bis zu welcher Konsequenz. Sie ist ebenfalls festgeschrieben und läuft eben nur digital ab, also JA- oder NEIN- Entscheidung.

3. Fuzzy-Logic, mehr als eine adaptive, automatische Kontrolle und Regelung

3.1. Komplexes Problemfeld allgemein

In VDI 3402 sind die Einstell- und Arbeitskennwerte in der Funkenerosion definiert. Im wesentlichen sind dies der Arbeitsstrom, der Entladestrom, die Arbeitsspannung, die Entladespannung, die Leerlaufspannung, die Zündspannung, die Impulsfrequenz, die Entladefrequenz, das Tastverhältnis, das Entladeverhältnis, das Frequenzverhältnis u.ä..

3.2. Auswahl der Teilbereiche

Die Abtragsrate, der relative Verschleiß der Arbeitselektrode und die erzielbare Oberflächenrauhigkeit stehen in direkter Abhängigkeit zur Impulsdauer. Oder, durch die Stromstärke, die Pausendauer und die Impulsdauer wird der Verschleiß und der Abtrag beeinflußt. Die Art der Bearbeitung (einfacher Senkvorgang oder komplizierter Planetarablauf), das Elektrodenmaterial, das Material des zu bearbeitenden Werkstückes, die zu bearbeitende Fläche sowie die Art der Spülung mit zusätzlichen programmierbaren Düsen haben ebenfalls starken Einfluß auf das Erodierergebnis.

3.2.1. Zustand im Funkenspalt

Der Meßwert, abhängig von den ausgewählten Technologieparametern ist der momentane, tatsächliche Zustand (Verschmutzungsgrad) im Funkenspalt. Bei zunehmender Erodiertiefe und entsprechend schlechter Einflußnahme durch externe Spülungsmethoden (Intervallspülung; pulsierendes Abheben u. ä.) stets verändert, verhält sich dieser Meßwert sehr fließend, ja verschwommen, nicht klar definiert. Innerhalb kürzester Zeit kann der Zustand ein wenig besser, oder ein wenig schlechter oder aber einfach ganz anders.

3.2.2. Änderung bestimmter Parameter

Manchmal ist die Impulspause z.B. ein bißchen zu klein, manchmal ein bißchen zu groß. Auch kommt es vor, daß bei einem sog. Timer-Betrieb der Abheberweg oder auch die Rückzug- und Zustellfrequenz etwas zu klein oder etwas zu groß gewählt wurde. Diese Parameter sind es, die immer Einfluß auf einen Erodiervorgang haben und bei einem wirtschaftlichen Einsatz der Erodieranlage nicht gleichmäßig ablaufen sollten.

3.2.3. Abhängigkeit der Parameter

Wie bereits oben genannt, stehen sehr viele dieser Einstellparameter in direkter Abhangigkeit. Die Veränderung nur einer Größe zieht eine Vielzahl von Konsequenzen nach sich. Und nie ist bei gleichen oder ähnlichen Erodierarbeiten exakt die gleiche Konsequenz zu ziehen.

3.2.4. Regelverhalten

Das Regelverhalten kann also nur verschwommen oder unklar definiert werden. Genau hier hat eine Fuzzy-Logic die entscheidenden Vorteile gegenüber jeder automatischen oder adaptiven Regelung.

Nur der erfahrene Erodierer ist in der Lage, eventuell noch konsequenter als diese Fuzzy-Logic eine moderne Erosionsanlage zu regeln.

Allerdings ist das nur möglich, wenn dieser erfahrene Erodierer ständig an der Anlage steht und den Erodierprozess auch ständig beobachtet.

4. Konsequenz

4.2. Ergebnis am Werkstück

4.1.1. Leistung qualitativ

Ein sog. "Clogging", nacherodieren an einer Stelle durch angesammelten Abtragschlamm, wird auszuschließen sein, besonders bei schmalen, tiefen Einsenkungen, wie z.B. Rippen, Stege. Ebenso sind Beschädigungen an Elektrode oder Werkstück durch Lichtbogen oder Verschmelzungen auszuschließen. Dies ist besonders kritisch bei rechtwinkeligen Hinterschnitten und tiefen Einsenkungen in einem stark chromhaltigen Material.

4.1.2. Leistung zeitlich

Anhand von ausgiebigen Testreihen konnten wir bei den unterschiedlichsten Senkerodieraufgaben ausschließlich Leistungsverbesserungen durch Einsatz der Fuzzy-Logic nachweisen. Selbst bei Aufgaben die im normalen Erodiervorgang unter 1 Stunde Dauer liegen, konnte die Zeit um nochmals 15 Minuten verbessert werden.

4.2. Wirtschaftlichkeit

In den meisten Fällen ist durch den Einsatz der Fuzzy-Logic die zu erbringende Vorarbeit geringer, da durch das sehr schonend mögliche Erodieren die Verschleißraten der Elektroden minimal sind. Meist ist eine einzige Elektrode ausreichend, um zu dem gewünschten Ergebnis im Werkstück zu kommen.

Dadurch sind neben den reinen Kosten für die Materialbeschaffung und Erstellung der Elektroden auch Personal- und Maschinenkosten für die anderen vorbereitenden Fertigungen einzusparen.

5. Auswirkungen auf den Werkzeug- und Formenbau

5.1. Betrieb der Anlage

Der Betrieb der Anlagen ist weitestgehend vollautomatisch, da komplette Steuerungs- und Regelungseinrichtungen in diesen Anlagen Standard sind. Selbst Werkzeug- und Werkstückwechsel-Einrichtungen sind im Fuzzy- Master-Concept realisiert.

5.2. Personaleinsatz

Ein zusätzliches, positives Merkmal der Fuzzy- Logic ergibt sich aus den ständig, während des Betriebes der Anlage veränderten Parametern. Diese werden von der Regeleinrichtung Fuzzy-Logic ständig visuell angezeigt und erlauben somit auch einem nicht so erfahrenen Bediener der Anlage einen besonderen Lerneffekt.

5.3. Nutzen für den Betrieb

Die Verfügbarkeit dieser Anlagen wird somit wesentlich verbessert, da auch in Zeiten der Abwesenheit des erfahrenen Bedienpersonals diese Anlagen besser genutzt werden können. Daraus ergibt sich auch ein sehr günstiges Leistungsverhältnis bei der Kalkulation für den Nutzungsgrad und den Maschinenstunden-Satz.

6. Schlußbetrachtung

Der Einsatz dieser modernsten Erodiertechnologie ermöglicht eine 30 - 50 % höhere Effizienz. Sie ist richtungsweisend in der konsequenten Nutzung modernster Regelungstechnik im Bereich von Werkzeugmaschinen.

Die wesentlich reduzierten Durchlaufzeiten und der hohe Nutzungsgrad geben auch dem Klein- und Mittelbetrieb optimale Voraussetzungen zu kostengünstiger Fertigung, bei gleichzeitiger Qualitätsverbesserung.

Mit der kontinuierlichen Weiterentwicklung dieses Expertenkonzeptes, der Fuzzy-Logic in der Funkenerosionstechnologie, ist MITSUBISHI erstmalig einen Schritt hin zur künstlichen Intelligenz bei Steuerungen im Bereich Universal-Werkzeugmaschinen gegangen.

Fuzzy Simulation
thermischer Trennverfahren

A. Kraslawski
Department of Chemical Technology
Lappeenranta University of Technology
P.O.Box 20
SF-53851 Lappeenranta/Finland

A. Gorak
Fachbereich Chemietechnik
Lehrstuhl für Thermische Verfahrenstechnik
der Universität Dortmund
Postfach 500 500
D-4600 Dortmund 50

1 Einleitung

Bei der Herstellung fast aller Erzeugnisse entstehen Gemische mehrerer Komponenten, die im Regelfall wieder getrennt werden müssen. Unsere industrialisierte Gesellschaft könnte nicht funktionieren ohne eine breite Palette von Trennverfahren, die an den verschiedensten Stellen des Produktionsprozesses eingesetzt werden. Thermische Trennverfahren zählen zu der Trennmethoden, die am häufigsten in der chemischen Industrie angewandt werden. Sie erfordern allerdings einen großen energetischen und apparativen Aufwand. So liegt zum Beispiel in den USA der Anteil der Rektifikation am gesamten Energieverbrauch bei 3%. Das ist mehr als für die ganze amerikanische Flugzeugindustrie notwendig ist. Eine geringe Energieeinsparung bei der Rektitfikation kann also beträchtliche Gewinne bringen. Dies ist u.a. mittels der Rechnersimulation möglich, die als Analyse der mathematischen Prozeßmodelle bezeichnet werden kann. Sie kann die ohnehin teuren Experimente ergänzen oder teilweise ersetzen.

Zu den meist verbreiteten thermischen Trennverfahren zählt die Rektifikation, deren Ziel es ist, flüssige Gemische in möglichst reine Komponenten zu trennen. Die Anforderungen an die Reinheit sind allerdings oft nur in einer verbalen Form formuliert (z. B. "der Ethanolgehalt im Vodka muß so hoch sein, daß das Produkt gut schmeckt"). Wie man - mit Hilfe von Fuzzy-Methoden - solche

Kriterien bei der Rechnersimulation von Rektifizierprozessen berücksichtigen kann, wird im ersten Rechenbeispiel gezeigt.

Ziel jeder Simulationsmethode eines thermischen Trennverfahrens ist es, die Trenneffekte vorauszuberechnen, die sich aus einer bestimmten Anlagenkonfiguration ergeben. Die Ein- und Ausgabedaten sind in einer scharfen Form definiert, da alle auf dem Markt verfügbaren Simulatoren nur mit diesem Zahlentyp arbeiten können. Im zweiten Rechenbeispiel wird es gezeigt, wie man mit Hilfe eines scharfen Simulators aus unscharfen Eingabedaten unscharfe Schlüsse ziehen kann.

2 Mathematische Modellierung einer Rektifizierkolonne

Bei einer Rektifikation handelt es sich im allgemeinen um die Trennung von zwei oder mehr Komponenten. Die Komponenten müssen allerdings in der Regel als Flüssigkeit vorliegen und unterschiedliche Siedetemperaturen aufweisen. Durch eine Wärmezufuhr erreicht man einen Siedezustand der zu trennenden Mischung. Dabei ist der entstehende Dampf reicher an der leichtersiedenden Komponente als die siedende Flüssigkeit. Solch ein Prozeß wird als Destillation bezeichnet /1/. Die Rektifikation ist - vereinfachend gesagt - ein vielfacher Siede- und Kondensationsprozeß, wodurch der Dampf immer weiter mit der leichtersiedenden Komponenten anreichert und in der Flüssigkeit immer mehr schwersiedende Stoffe zurückbleiben.

Die Rektifikation wird in Kolonnen durchgeführt, in welchen die zu trennende Zulaufmischung etwa in der Kolonnenmitte eingespeist wird. Aufgrund der dem Kolonnensumpf zugeführten Wärme verdampft die Flüssigkeit und steigt als Dampf aufwärts, wobei der Anteil der leichtersiedenden Komponente im Dampf umso größer ist, je höher sich der Dampf in der Kolonne befindet. Das Destillat besteht oft nur aus den leichten Komponenten, der Sumpf aus den schweren.

Die Aufgabe eines Ingenieurs ist es, die Höhe der Kolonne so zu bestimmen, daß die Destillatzusammensetzung der Anforderungen der Kunden entspricht. Für die Vorausberechnung dieser Höhe benutzt man mathematische Modelle, die den stationären oder dynamischen Zustand einer Rektifizierkolonne beschreiben.

Wie wird ein mathematisches Modell eines thermischen Trennprozesses formuliert? Die Antwort auf diese Frage versuchen wir am Beispiel einer Rechnersimulation für eine Kolonne zur Mehrstoffrektifikation zu geben. Die Kolonne wird in eine bestimmte Anzahl von miteinander verbundenen Elementen unterteilt (Bild 1). Dadurch entsteht ein physikalisches Modell der Kolonne, in welchem angenommen wird, daß in jedem Element eine sprunghafte Änderung des Zustandes von flüssiger und gasförmiger Phase stattfindet. Solche Elemente werden auch als Stufen bezeichnet.

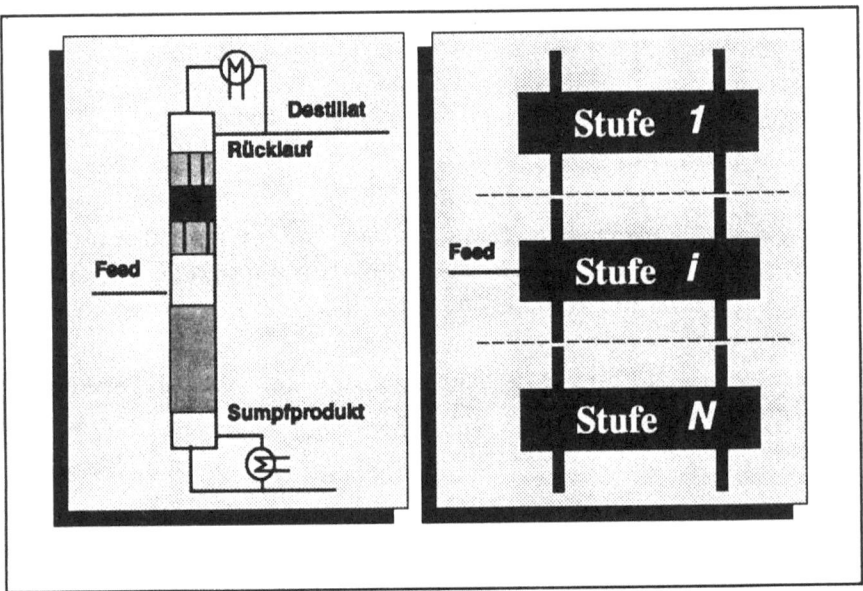

Bild 1 Stufenmodell einer Rektifizierkolonne

In der Verfahrenstechnik ist das "Modell der theoretischen Stufe" das am häufigsten angewandt. Es basiert auf der Grundlage eines theoretischen Bodens /1/, wobei hier unterstellt wird, daß die aus dieser Stufe abfließenden Phasen sich im Gleichgewichtszustand befinden. Wie weit der in der Praxis verlaufende Prozeß von diesem idealen Zustand abweicht, wird durch den im Grunde genommen künstlich eingeführten Begriff des Bodenverstärkungsverhältnisses beschrieben. Der ganze Prozeß wird als Stufen-kaskade dargestellt, in welcher jede Stufe mit der nächsten durch die Massen und Energieströme verbunden ist.

Aufgrund solcher physikalischen Darstellung läßt sich dann ein mathematisches Modell der Kolonne schreiben. Die Lösung eines solchen Modells erlaubt es, die Konzentrationen jeder Komponente auf einer beliebigen Kolonnenhöhe (also ein Konzentrationsprofil) zu bestimmen. Die Konzentration eines Kopfproduktes soll der geforderten Destillatzusammensetzung entsprechen. Sie kann nur bei einer bestimmten Kolonnenhöhe erreicht werden, die sich aus den Simulationsrechnungen ergibt.

3 Optimale Kolonnenhöhe bei unscharfen Anforderungen an die Produktreinheit

Der Projektant einer solchen Kolonne ist oft mit unscharfen Kriterien bezüglich der Produktreinheit konfrontiert. Der Kunde bestimmt z. B. nur die Wichtigkeitshierarchie der Komponenten oder verlangt, daß die Konzentration eines Produktes "möglichst nah am gewünschten Wert" liegt. Da die Kunden oft widersprüchliche Wünsche haben, entsteht ein Problem der Mehrkriterienoptimierung, in welchem die Kriterien in einer unscharfen Form definiert werden /2,3/. Gesucht wird nach einer Kolonnenhöhe, die die Erfüllung möglichst vieler Wünsche garantiert. Die Kolonnenhöhe muß selbstverständlich als eine scharfe Zahl vorliegen.

Das Problem wird anhand folgenden Beispiels illustriert: in einer Trennkolonne (Bild 2) soll ein Gemisch aus vier Komponenten getrennt werden und zwar eine Mischung von Propan, Butan, Pentan und Hexan mit der Zusammensetzung von 38%, 4%, 36% und 22%. Der von unterschiedlichen Kunden gewünschte Gehalt an jeder Komponente im Destillat beträgt: 50%, 25%, 10% und 2%. Aufgrund einer Rechnersimulation können wir die Konzentrationsprofile entlang der Kolonne berechnen. Wir stellen fest, daß leider keine Kolonnenhöhe existiert, auf welcher das Destillat die geforderte Zusammensetzung hat. Die Propankonzentration von 50% wird bei einer Kolonnenhöhe von 5,7 m erreicht. Für diese Höhe ergeben sich aber keinesfalls die Werte, die dem geforderten Gehalt der anderen Komponenten entsprechen (Bild 3).

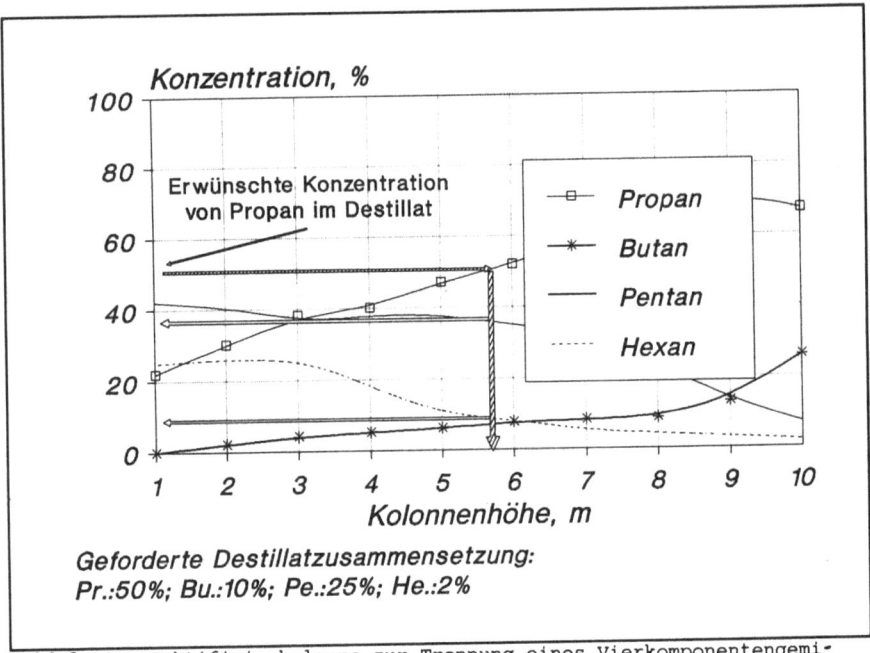

Bild 2 Eine Rektifizierkolonne zur Trennung eines Vierkomponentengemisches

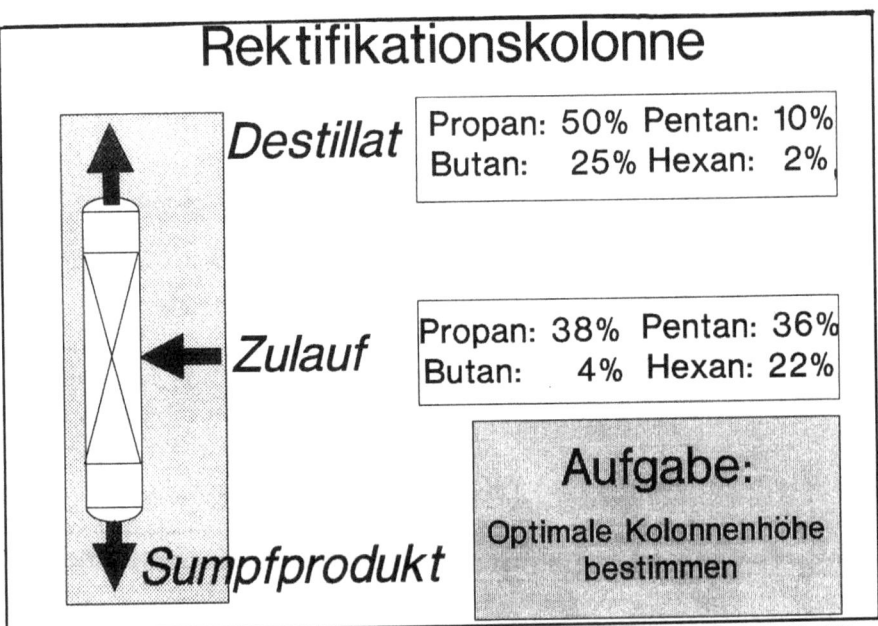

Bild 3 Konzentrationsprofil in einer Rektifizierkolonne zur Trennung von Kohlenwasserstoffen

Das Problem wird in folgender Weise gelöst: man bestimmt eine unscharf definierte Destillatzusammensetzung, die sich aus den Wünschen von Kunden ergibt. Bild 4 zeigt den Verlauf einer Zugehörigkeitsfunktion für die leichteste Komponente.

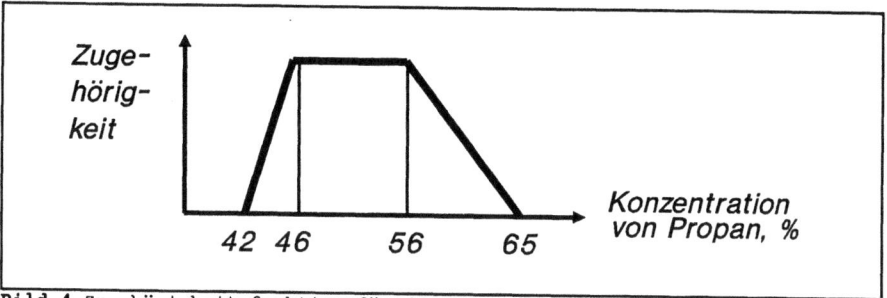

Bild 4 Zugehörigkeitsfunktion für die erwartete Propankonzentration im Destillat

Man befragt zusätzlich die Kunden, in welcher Wichtigkeitshierchie die Komponenten einzuordnen wären. Danach bildet man ein Optimierungskriterium in einer Form:

$$d = \sum_{i=1}^{n} w_i |x(h) - x_i^D|$$

wobei

w_i - eine Wichtung für jede Komponente i,
x_i^D - eine unscharfe Zusammensetzung des geforderten Destillates,
$x(h)$ - Flüssigkeitszusammensetzung auf der Höhe h, die sich aus den Simulationsrechnungen ergibt.

Anschließend berechnet man das Kriterium für die Höhe h in einem technisch sinnvollem Bereich. Die Höhe, für welche dieses Kriterium den kleinsten Wert erreicht, entspricht der Kolonnengröße, für welche die Wünsche möglichst aller Kunden erfüllt wurden. In dem Rechenbeispiel ist es die Höhe von 9,7 m, die eine Destillatzusammensetzung von 62%, 20%, 5% und 13% garantiert (Bild 5).

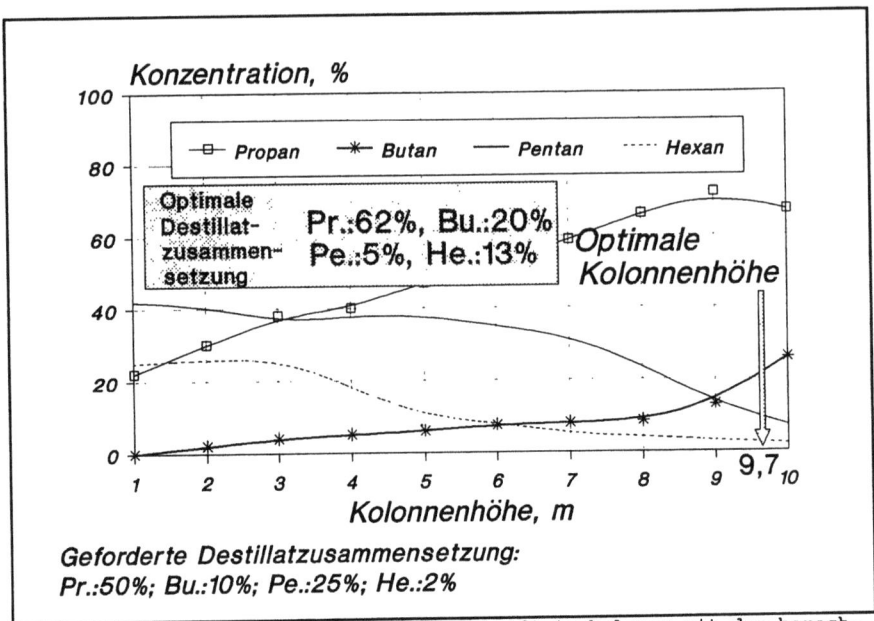

Bild 5 Konzentrationsprofile in einer Rektifizierkolonne mit der berechneten Kolonnenhöhe und optimaler Destillatzusammensetzung

4 Erfassung der Unschärfe von Modellparametern durch die Vertex-Methode

Das oben aufgeführte Beispiel illustriert, wie man bei der Simulation aufgrund unscharfer Kriterien scharfe Schlüsse ziehen kann. Aber was kann man tun, wenn man unscharfe Ein- und Ausgangsdaten haben möchte, aber nur über einen mit scharfen Zahlen arbeitenden Prozeßsimulator verfügt?

Um das Problem definieren zu können, müssen einige Begriffe eingeführt werden. Bei jeder Simulation einer Rektifizierkolonne muß der Ingenieur einige Parameter festsetzen, die als Grundlage für die Auslegungsarbeiten dienen. Die Parameter können in zwei Gruppen unterteilt werden:

- Betriebsparameter, wie z. B. die Menge der zugeführten Flüssigkeit, des abgenommenen Destillates oder die notwendige Wärmemenge,
- Modellparameter, wie z. B. das am Anfang erwähnte Bodenverstärkungsverhältnis, das die Effizienz der Kolonne beschreibt.

Nachdem alle Parameter vom Projektanten bestimmt worden sind, kann man die Produktzusammensetzung mit einem Prozeßsimulator berechnen. In der technischen Realität schwanken aber die Parameter immer um einen Sollwert, so daß sie oft nur in einer unscharfen Form definiert werden können. Eine Vertex-Methode erlaubt uns, aufgrund der unscharfen Eingangsdaten die unscharfen Antworten zu bekommen und einem scharfen Simulator dabei zu benutzen.

Die Anwendung dieser Methode wird im folgenden Beispiel illustriert: einer Rektifizierkolonne wird ein flüssiger Zulauf zugespeist, dessen Durchfluß prinzipiell 70 kg/s betragen soll, aber auch in gewissen Grenzen schwanken kann. Der Bodenverstärkungsverhältnis beträgt normalerweise rund 80 %. Aufgrund der Erfahrung eines (oder mehrerer) Anlagefahrers können die Abweichungen in Form einer Zugehörigkeitsfunktion dargestellt werden (Bild 6). Die Frage ist, wie der Gehalt einer Wertkomponente A im Destillat schwankt.

Das Problem wird mit der Vertex-Methode gelöst /4/, die die Datenunsicherheit direkt zu untersuchen erlaubt. Die Vertex-Methode besteht aus folgenden Schritten:

Zuerst werden die Zugehörigkeitsfunktionen der Eingangsdaten definiert. Sie ergeben sich entweder aus der Erfahrung eines Projektanten (Modelldaten) oder des Anlagefahrers (Betriebsdaten). Zudem wird die Niveaumenge der unscharfen Eingangsdaten auf unterschiedlicher α-Höhe bestimmt. Dadurch entstehen die Schnittpunkte der α-Niveaumengen von Eingagsdaten, für welche die Ergebnisse mit einem "scharfen" Simulator berechnet werden (Bild 7). Aufgrund der Analyse der für die Schnittpunkte berechneten Ausgangsdaten läßt sich anschließend die Zugehörigkeitsfunktion von Ergebnissen bestimmen. Somit wird eine fuzzy-Form der Konzentration jeder Komponente im Destillat, also auch des Gehaltes von Komponente A konstruiert (Bild 8). Die Vertex-Methode ist auf beliebige Parameteranzahlen auszudehnen.

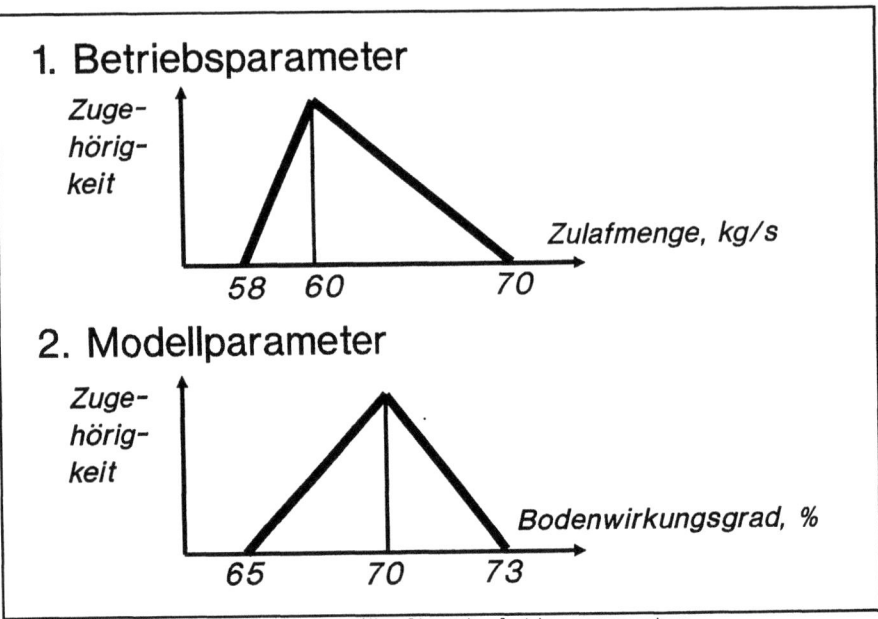

Bild 6 Zugehörigkeitsfunktion für die Simulationsparameter

5 Schlußbemerkung

In der Industrie verwendet man tagtäglich Rechenprogramme, die unterschiedliche verfahrenstechnische Prozesse simulieren. Alle diese Simulatoren können nur scharfe Zahlen bearbeiten. In diesem Beitrag wurde die Anwendung der Fuzzy-Simulation für die Erfassung unscharfer Produktzusammensetzungen oder Modellparameter anhand von zwei Beispielen gezeigt. Da die Fuzzy-Methoden der klassischen Sensitivitätsanalyse überlegen sind /4/, werden sie sicherlich in der Zukunft bei der Auslegung von verfahrenstechnischen Anlage große Akzeptanz finden.

6 Literaturverzeichnis

1. Schlünder E.M.: "Destillation, Absorption, Extraktion", Thieme Verlag, Stuttgart (1989)
2. Kraslawski. A., Gorak A., Vogelpohl A.: Determined and fuzzy criteria in multicomponent distillation calculations, Proc. CEF, Giardini Naxos, April (1986)
3. Kraslawski A., Gorak A.: Optimization of a Distillation Column: Fuzzy Multiobjective Approach, I.Chem.Symposium Series No. 104 (1989)
4. Kraslawski A., Gorak A.: Anwendung der Fuzzy-Logik in der Simulation thermischer Trennverfahren, Jahrestreffen der Verfahrensingenieure, Wien (1992)

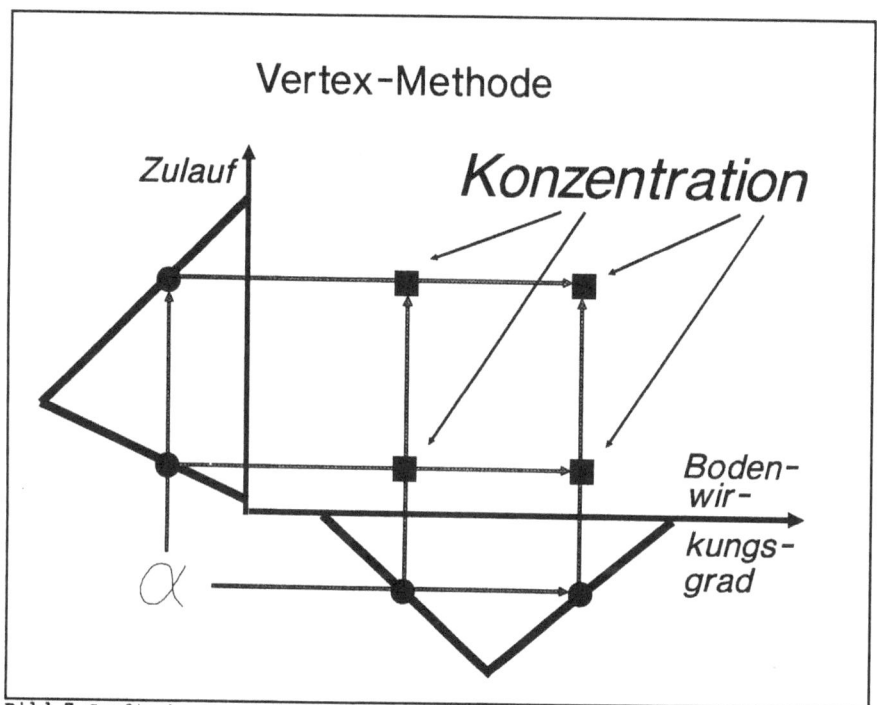

Bild 7 Grafische Darstellung der Vertex-Methode für zwei unscharfe Modellparameter

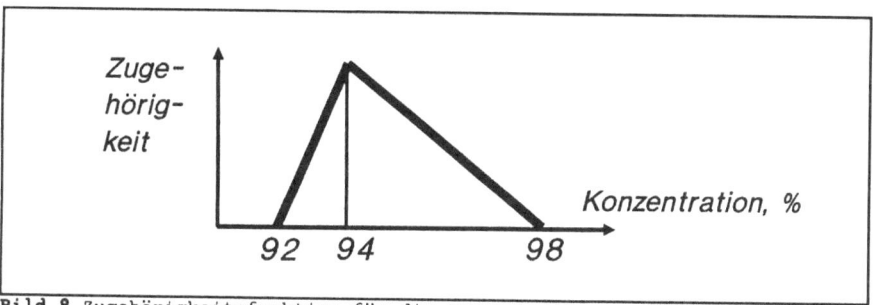

Bild 8 Zugehörigkeitsfunktion für die Konzentration von einer wertvollen Komponente im Destillat

Anwendung von Fuzzy-Control in der thermischen Verfahrenstechnik

Dieter Schulz
Ingenieurbüro
Am Heisterbach 83
4600 Dortmund 30

1 Einleitung

Die Entwicklung von Regelungen für Aufgabenstellungen in der thermischen Verfahrenstechnik ist besonders schwierig, da die hier zu findenden Systeme im allgemeinen nichtlinear sind, über mehrere Einflußgrößen verfügen und die Prozesse räumlich verteilt sind. Hinzu kommen wechselnde Lasten und ungenaue dynamische Prozeßmodelle. Bei diesem Aufgabengebiet hat sich der Einsatz von Fuzzy-Control besonders bewährt. Hier ist es nicht nur interessant, die Regelungen empirisch an den echten Strecken zu entwickeln, sondern es ist in vielen Situationen vorteilhaft die Fuzzy-Logic an simulierten Modellen zu entwerfen. In diesem Beitrag sollen zunächst kurz die Besonderheiten der Aufgaben in der thermischen Verfahrenstechnik besprochen werden. Anschließend wird die Grundstruktur von Einsatzmöglichkeiten der Fuzzy-Control wiedergegeben und die Vorgehensweise bei der Entwicklung von Fuzzy-Control Systemen in der Verfahrenstechnik diskutiert.

1. Aufgabenstellungen in der thermischen Verfahrenstechnik

Die meisten Aufgabenstellungen der thermischen Verfahrenstechnik lassen sich auf die Grundprinzipien der Wärmeübertragung und der thermischen Trennverfahren zurückführen. Um die Prozesse, die Gegenstand der thermischen Verfahrenstechnik sind, besser charakterisieren zu können, wird zunächst kurz auf die physikalischen Grundlagen eingegangen.

1.1 Wärmeübertragung

Die Wärmeübertragung in der thermischen Verfahrenstechnik beruht auf drei verschiedenen physikalischen Prinzipien. Das erste Prinzip (Abb. 1) ist die Wärmeleitung. Die Wärmeleitung behandelt den Durchgang eines Wärmeflusses durch einen Festkörper auf Grund einer Temperaturdifferenz. Die zugrunde liegenden Differentialgleichungen sind in diesem Fall linear und einfach zu behandeln. Der Wärmefluß ist proportional zur Temperaturdifferenz, zur Fläche und zur Wärmedurchgangszahl.

Wesentlich schwieriger stellt sich das Phänomen der Konvektion (Abb. 2) dar. Man muß sie berücksichtigen, um den Wärmeübergang von der Oberfläche eines Festkörpers in Luft oder ein flüssiges Medium zu berechnen. Die genaue Größe der Konvektion hängt von einer Reihe von Randbedingungen ab. Diese kann man über die Benutzung der Nusseltzahl, einer charakteristischen Länge l und der Wärmeleitzahl berechnen. Hieraus ergibt sich die Wärmedurchgangszahl k. Die Geometrie eines Wärmetauschers und die Strömungsgeschwindigkeit können die Größe der Zahl k stark beeinflussen.

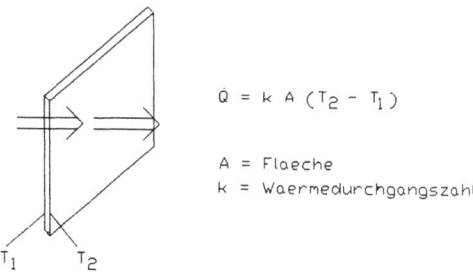

Abb.1: Wärmeübertragung durch Wärmeleitung durch eine Platte

Das dritte physikalische Prinzip zur Übertragung von Wärme zwischen zwei Flächen ist die Strahlung. Die Übertragung von Strahlungswärme ist stark nichtlinear, da die übertragenen Leistung proportional zur vierten Potenz der Oberflächentemperaturen ist (siehe Abb. 3).

In realen technischen Systemen sind alle drei Mechanismen an der Wärmeleitung beteiligt. Die gesamte systemdynamische Charakteristik eines wärmeübertragenden Systems erhält man durch Parallelschaltung oder Kaskadierung der einzelnen Effekte. Es ist leicht einzusehen, daß bei der Wärmeübertragung ein nichtlineares System höherer Ordnung entsteht, dessen einzelne Größen in vielen Fällen nicht exakt bestimmbar sind.

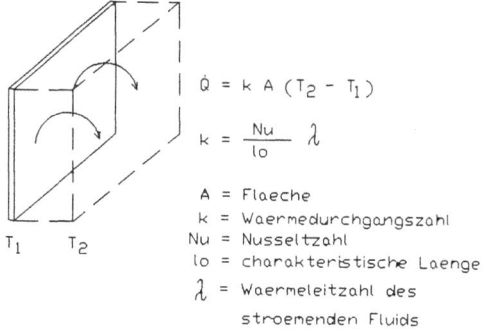

Abb.2: Wärmeübertragung durch Konvektion an einer Oberfläche

In Anlagen kommen diese physikalischen Effekte in Wärmetauschern und beim Heizen und Kühlen von Reaktionen zum Tragen. Gleichgültig, ob Wärmetauscher im Gleich- oder Gegenstromprinzip arbeiten, treten hier einige spezielle Probleme auf. Die Größe des sich einstellenden Wärmestroms ist nicht nur abhängig von der Kühl- und Wärmemitteltemperatur, sondern auch strömungsabhängig.

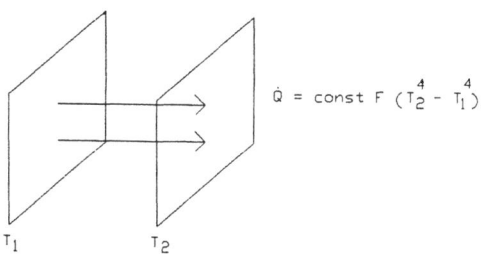

Abb.3: Wärmeübertragung durch Strahlung zwischen zwei Oberflächen

Treten zusätzlich Änderungen des Aggregatzustandes des Kühl- oder Wärmemittels auf, ist der quantitative Zusammenhang stark nichtlinear.

Beim Heizen und Kühlen von Reaktionen unterscheidet man zwischen Mantelkühlung und Innenkühlung. Besitzt ein System eine Mantelkühlung, so ist dessen thermodynamische Auslegung wesentlich schwieriger, da das Kühlmittel einen schwächeren Kontakt mit dem zu kühlenden Medium hat. Die hier auftretenden Probleme sind gleich denen in einem Wärmetauscher. Zusätzlich tritt die Problematik der Reaktionen hinzu. Die Reaktion des chemischen Systems kann exotherm oder endotherm sein. Die Stärke der Reaktion ist bei kritischen Punkten oft stark temperaturabhängig. In vielen Fällen ist es notwendig, je nach Temperaturbereich der Reaktion, zu Kühlen oder zu Heizen.

1.2 Trennverfahren

Thermische Trennverfahren dienen dazu, Stoffmischungen, die in unterschiedlichen Phasen vorliegen können, zu trennen. Voraussetzung ist, daß bei gleichzeitig nebeneinanderstehenden Phasen, die Stoffmischungen unterschiedliche Zusammensetzungen besitzen. Die entsprechenden Phasenübergänge werden durch Erhitzen oder Kühlen hervorgerufen. Beispiele für thermische Trennverfahren sind Vorgänge, die auf Kondensation oder Destillation von Substanzen beruhen. Die wichtigsten thermischen Trennverfahren sind folgende technische Prozesse:
- Destillationskolonnen,
- Rektifikationskolonnen,
- Verflüssigungsanlagen,
- Trocknungsanlagen und
- Ionenaustauschanlagen.

2 Allgemeine Probleme der thermischen Verfahrenstechnik

Sieht man sich die genannten technischen Prozesse vom regelungstechnischen Standpunkt aus an, findet man, daß bei allen Prozessen ähnliche Probleme auftreten. Die meisten der genannten technischen Probleme sind nichtlinear. In den Fällen, in denen Phasenübergänge auftreten, findet man sogar stark nichtlineares Verhalten. Außerdem sind die technischen Prozesse dadurch gekennzeichnet, daß sie in der Regel mehrere Einflußgrößen besitzen. Das heißt, die zu regelnde Größe wird von mehreren Größen des technischen Prozesses stark beeinflußt. Hinzu kommt, daß ein Teil der Prozesse als räumlich verteilt angesehen werden muß. Dies ist sowohl bei Wärmetauscher, als auch bei Reaktoren der Fall. Dadurch entstehen Schwierigkeiten in der Analyse der Prozesse und bei dem Entwurf von entsprechenden Regelungen. Wechselnde Lasten haben im allgemeinen einen starken Einfluß auf das Verhalten des Gesamtsystems. Die zu entwerfenden Regelungen müssen nicht nur auf rein regelungstechnische Gesichtspunkte, wie Stabilität und Präzision optimiert werden, sondern es sind auch betriebliche Gesichtspunkte zu beachten wie zum Beispiel, die energieoptimale Führung eines Prozesses. Einige der eben genannten Punkte führen dazu, daß nur ungenaue thermische Prozeßmodelle entwickelt werden können.

3 Vorteile von Fuzzy-Control

Auf Grund der Eigenschaften der Prozesse in der thermischen Verfahrenstechnik bestehen erhebliche Probleme bei dem Entwurf konventioneller Regelalgorithmen. Werden die Regelalgorithmen nach theoretisch regelungstechnischen Verfahren entworfen, hat man die Schwierigkeit, daß auf der einen Seite nur ungenaue Prozeßmodelle zur Verfügung stehen und sich auf der anderen Seite die entsprechenden Entwurfsverfahren nur schwierig auf die hier zu findenden nichtlinearen Systeme anwenden lassen. Geht man mehr von der praktischen Seite an die Lösung der Probleme heran, stellt man bald fest, daß die in der praktischen Regelungstechnik eingesetzten PID-Regler in diesem Gebiet oft nur schwer einsetzbar sind. In vielen Fällen ist es notwendig, die Parametrierung eines solchen PID-Reglers betriebspunktabhängig zu fahren.

Werden dagegen Systeme basierend auf Fuzzy-Control-Algorithmen entwickelt, gibt es zwei Vorteile. Der eine Vorteil liegt im Entwurfsprozeß begründet. Hier ist es möglich, die Entwicklung der Regelung im Zeitbereich durchzuführen. Man bewegt sich bei der Anwendung von Fuzzy-Control auf einer Ebene, die dem Erfahrungshorizont eines Technikers oder Ingenieurs näher kommt. Wird die Entwicklung von Fuzzy-Control online eingesetzt, kann man begründet auf der Erfahrung einen Regelalgorithmus entwerfen und an den einzelnen Betriebspunkten stückweise optimieren.

Der andere Vorteil der Fuzzy-Logic liegt darin begründet, daß Fuzzy-Logic-Systeme von ihrer Natur her nichtlinear sind. Bei geeigneter Anpassung der Nichtlinearitäten des Fuzzy-Control-Systems an die Nichtlinearität der zu regelnden Strecke kann hier ein robusteres Verhalten erreicht werden. Diese Möglichkeiten erschließen sich, wenn ein Fuzzy-Control-Regler eingesetzt wird. Weiterhin bietet Fuzzy-Control die Möglichkeit, regelungstechnische Systeme mit Erfahrungswissen zu kombinieren. Das Erfahrungswissen kann zum Beispiel die Erzeugung von Führungsgrößen je nach betrieblicher Situation ermöglichen oder es führt zur Einstellung von Reglerparametern in Abhängigkeit von mehreren technischen Größen eines Systems. Diese Verbindung von Erfahrungswissen mit Regelungen kann erstmalig mit der Fuzzy-Logic realisiert werden. Auf diese Weise ist es sowohl möglich, Vorgaben für Fuzzy-Control-Regler, als auch für konventionelle Regler zu erzeugen und zu implementieren.

In den letzten Jahren stand bei der Entwicklung von Fuzzy-Control die empirische Entwicklung am echten System im Vordergrund. Es soll nachfolgend dargestellt werden, daß die Entwicklung auch an simulierten Prozessen, die halbempirischen Charakter haben können, durchführbar sind. Ist eine solche Entwicklung möglich, ergibt sich unter Umständen ein großer Vorteil bei der Entwicklung des Fuzzy-Control-Systems, da die Entwicklung unter Umständen noch schneller und mit noch geringeren Kosten durchgeführt werden kann.

4 Vorgehensweise bei der Entwicklung von Fuzzy-Control

Wie in dem letzten Kapitel angesprochen, ist es möglich, Fuzzy-Control-Systeme auch an simulierten Modellen zu entwickeln. Dies ist dann vorteilhaft, wenn die Arbeit an den Originalanlagen entweder mit Risiken oder mit einem erheblichen Materialaufwand verbunden ist. In der produzierenden Industrie ist oft eine Einstellung von Fuzzy-Reglern an echten Anlagen mit einem großen Zeitaufwand verbunden, wenn die Prozesse langsam sind und mit hohen Kosten, wenn das benötigte Material teuer ist. In vielen Fällen kann jedoch ein Prozeßmodell benutzt werden.

Die Entwicklung von Fuzzy-Control an Prozeßmodellen setzt jedoch spezielle Entwicklungssysteme voraus, die eine Verbindung von Fuzzy-Control mit der linearen und nichtlinearen Systemdynamik ermöglichen. Ein solches System ist das Entwicklungssystem AGO aus dem Ingenieurbüro Dr. Schulz in Dortmund. Hier hat man nicht nur die Möglichkeit, Streckendaten meßtechnisch zu erfassen, sondern auch die Fähigkeit, Parameter in Blockschaltungen für Streckenmodelle zu identifizieren. Dies trifft auch für Parameter nichtlinearer Streckenmodelle zu. Auf diese Weise können sehr einfache und praxisgerechte Modelle erzeugt werden, die den Entwicklungsprozeß wesentlich unterstützen. Einschränkend ist zu sagen, daß diese Vorgehensweise nur dann sinnvoll ist, wenn die Anzahl der bei dem Regelungsvorgang zu betrachtenden regelungstechnischen Größen nicht zu hoch ist. Bei mehr als zwei Eingangsgrößen des Systems stoßen die Möglichkeiten der nichtlinearen Prozeßidentifikation an Grenzen.

Ist diese Bedingung erfüllt, können das Streckenmodell und das Fuzzy-Logic-Modell in einer Blockschaltung entworfen und optimiert werden. Hier besteht die Möglichkeit, mit einer Optimierungsschale Parameter des Fuzzy-Logic-Systems anhand vorgegebener Gütekriterien im Zeitbereich optimal einzustellen.

Ziel des Entwicklungsprozesses ist ein generierter C-Quelltext, der das gesamte Automatisierungssystem, bestehend aus Fuzzy-Control und konventionellen Elementen beinhaltet. Zu erwähnen ist, daß eine solche Entwicklungsumgebung nicht nur die Integration von systemdynamischen Modellen mit der Fuzzy-Logic ermöglicht, sondern eine beliebige Kombination von Fuzzy-Logic-Elementen und konventionellen Elementen innerhalb einer regelungstechnischen Schaltung. Als konventionelle Elemente stehen lineare, nichtlineare, dynamische und steuerungstechnische Elemente zur Verfügung.

Das so erzeugte Automatisierungssystem muß in der Regel am Originalsystem noch fein optimiert werden. Der Aufwand für die Feinoptimierung ist jedoch wesentlich geringer als die empirische Entwicklung des Gesamtsystems.

5 Struktur von Fuzzy-Control-Systemen

5.1 Reine Fuzzy-Control-Systeme

Reine Fuzzy-Control-Systeme lehnen sich in ihrer Grundstruktur an klassische Reglerstrukturen an. Im Gegensatz zu diesen beruht ihre Wirkungsweise jedoch auf einer linguistischen Beschreibung von Nichtlinearitäten. Die Grundstruktur vieler Fuzzy-Control-Systeme ähnelt entweder dem PD- oder dem PI-Regler. Dazu ist es notwendig, zunächst die Regeldifferenz zu bilden und diese anschließend zu differenzieren, bzw. zu integrieren. Auf diese Weise erhält man den P-, I- und D-Anteil. In der Regel werden dann zwei dieser Anteile über eine Fuzzy-Logic-Schaltung miteinander verknüpft und erhält so den PD- oder PI-Anteil (siehe Abb. 4, 5). In dem eigentlichen Fuzzy-Logic-System wird zuerst eine Fuzzifizierung durchgeführt, indem

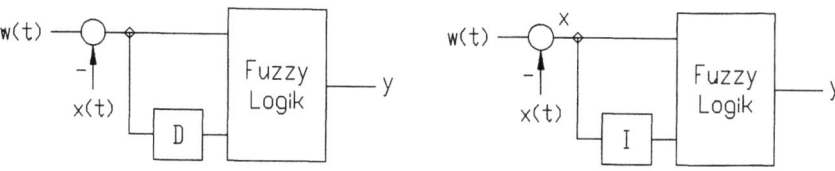

Abb.4: Schaltung eines PD-ähnlichen Fuzzy-Control-Systems

Abb.5: Schaltung eines PI-ähnlichen Fuzzy-Control-Systems

aus der Regeldifferenz und den daraus abgeleiteten Größen verschiedene Zugehörigkeitsfunktionen gebildet werden. Anschließend werden diese Größen über Fuzzy-Logic Regeln miteinander verknüpft. Aus diesen Größen wird dann eine Stellgröße y gebildet, indem die Ergebnisse der Regelwerke defuzzifiziert werden.

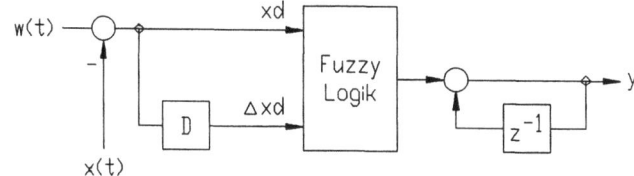

Abb.6: Fuzzy-Control-Systems, das einem PI-Regler nach dem Geschwindigkeitsalgorithmus entspricht

Von diesen Grundstrukturen gibt es in der Praxis verschiedene Varianten. Häufig ist die Variante, daß das Ausgangssignal des Fuzzy-Logic Werkes integriert wird, indem ein Akku bestehend aus einem Addierer und einem Analogspeicher hinter das Fuzzy-Logic-System geschaltet wird (Abb. 6). Auf diese Weise entspricht der gebildete Regler in seiner Grundstruktur einem digitalen Abtastregler nach dem Geschwindigkeitsalgorithmus. Das Ergebnis des Fuzzy-Systems ist eine Stellgrößenänderung. Bedingt durch das speichernde Glied am Ausgang wird aus dem P-Anteil am Eingang der Fuzzy-Logic Schaltung einen I-Anteil und aus dem D-Eingang ein P-Anteil eines Reglers. Diese Variante ist besonders dann vorteilhaft, wenn das Regelwerk Lücken hat. In einer weiteren Variante wird nicht die Regeldifferenz differenziert, sondern die Regelgröße selber. Dies wird dann eingesetzt, wenn das Fuzzy-Logic-System auf Störgrößenänderungen reagieren soll, aber nicht auf veränderte Führungsgrößen.

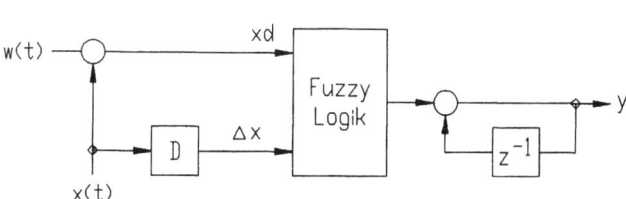

Abb.7: Variante des Fuzzy-Reglers aus Abbildung 6

5.2 Einstellung konventioneller Regler

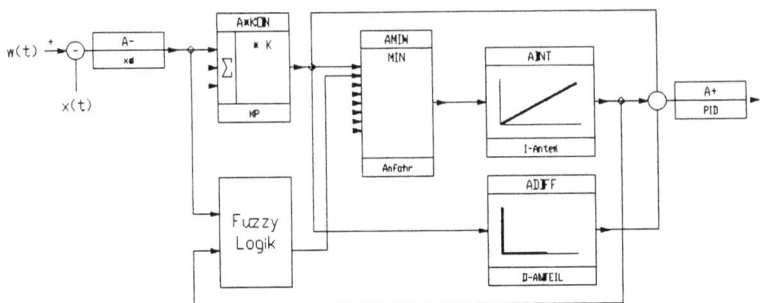

Abb.8: Konventioneller PID-Regler mit Fuzzy Anfahrschaltung

Fuzzy-Logic-Systeme können nicht nur isoliert eingesetzt werden, sondern auch zusammen mit konventionellen Regelungssystemen. Dies ermöglicht dem erfahrenen Regelungstechniker, seine Erfahrung in der konventionellen Regelungstechnik mit den Möglichkeiten der Fuzzy-Logic zu verbinden. Ein Anwendungsbeispiel liegt in der betriebspunktabhängigen Veränderung des I-Anteils. Bei konventionellen regelungstechnischen Schaltungen ist der I-Anteil auf der einen Seite eine notwendige strukturelle Komponente um einer langzeitigen Regeldifferenz entgegen zu wirken. Auf der anderen Seite verschlechtert ein hoher I-Anteil das Anfahren des Reglers und kann zu Instabilitäten bei bestimmten Streckentypen führen. Fuzzy-Logic-Systeme ermöglichen es, den I-Anteil in Abhängigkeit von dem Betriebspunkt einzustellen (siehe Abb.8). Hier gibt es verschiedene Varianten. In der Abbildung ist die Möglichkeit gezeigt, den I-Anteil in Abhängigkeit von der Regeldifferenz und dem augenblicklichen Wert des I-Gliedes einzustellen [2]. Diese Schaltung entspricht in etwa einer klassischen Anfahrschaltung, bietet jedoch durch die zwei Eingangsgrößen stärkere Möglichkeiten das I-Glied betriebspunktabhängig zu führen.

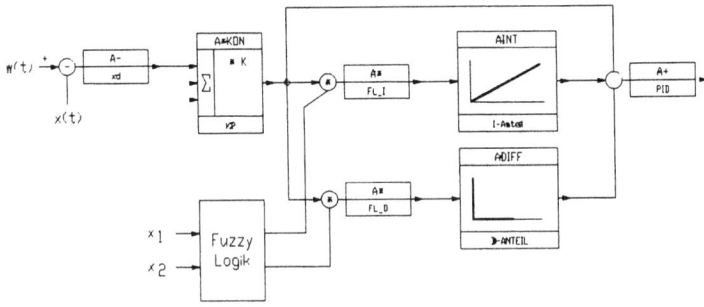

Abb.9: Konventioneller PID-Regler mit Fuzzy Adaption des I- und D-Anteils

Fuzzy-Logic Schaltungen können nicht nur einzelne Teile eines klassischen Reglers verändern, sondern sie können auch zu beliebigen Adaptionen regelungstechnischer Glieder dienen. Die Adaption des I- und D-Anteiles kann nicht nur auf Grund der Information der Regelgröße erfolgen, sondern auch auf Grund beliebiger Signale eines technischen Systems. Der Einsatz der Fuzzy-Logic hat hier den Vorteil, mehrere Eingangsgrößen auch nichtlinearer Systeme miteinander verrechnen zu können und sie so zu verarbeiten, daß ein stetiges Ausgangssignal in Abhängigkeit des Betriebspunktes eines regelungstechnischen Systems erreicht wird.

5.3 Sollwertvorgabe

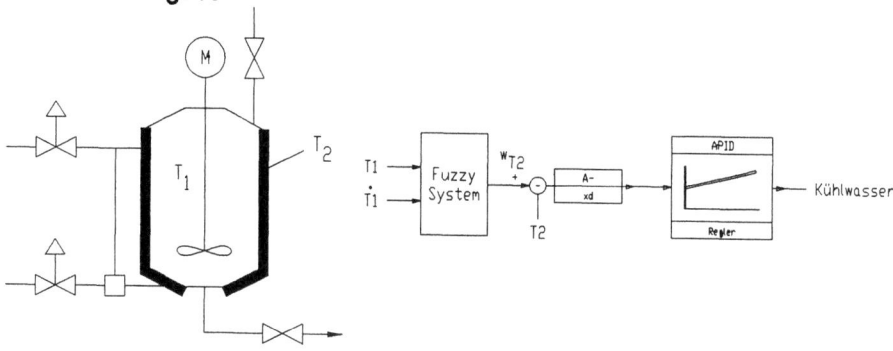

Abb.10: Mantelgekühlter Rühr-
kesselreaktor

Abb.11: Erzeugung der Führungsgröße T2

Eine weitere Möglichkeit ein Fuzzy-Logic-System mit einem konventionellen regelungstechnischen System zu kombinieren besteht in der Sollwertvorgabe für ein konventionelles Regelungssystem. Solche Sollwertvorgaben sind wichtig, wenn das Ziel einer Regelung von mehreren Eingangsgrößen abhängt. Solche Schaltungen lassen sich auch bei speziellen regelungstechnischen Kaskaden mit Gewinn einsetzen. In dem hier geschilderten Beispiel wird die Fuzzy-Logic für die Verbesserung der Kühlung eines Rührkesselreaktors mit einer Mantelkühlung eingesetzt. Rührkesselreaktoren mit Mantelkühlung sind relativ kritische Systeme, da die Kühlung einen relativ geringen Oberflächenkontakt des Kühlmittels mit dem zu kühlenden Medium besitzt. Die Entwicklung von Regelungen für solche Reaktoren mit exothermen Reaktionen sind besonders schwierig. Wie anfangs gezeigt, ist dieses System sehr stark temperaturabhängig und nichtlinear. Eine solche Regelung kann mit einem Fuzzy-Logic Anteil verbessert werden, indem aus der Abhängigkeit von der Innentemperatur des Reaktors und von der Ableitung der Temperatur eine Führungsgröße erzeugt wird [1]. Der PID-Regler regelt die Temperatur des Kühlmittels im Mantel des Reaktors. Dabei wird eine Temperaturdifferenz aus der vorgegebenen Temperatur für T2 und dem Istwert der aktuellen Temperatur im Mantel gebildet. Die Führungsgröße ist jedoch nicht konstant, sondern hängt von der Reaktion im Reaktor ab. Steigt zum Beispiel die Temperatur im Reaktor schnell oder ist die Temperatur im Reaktor hoch, muß stärker gekühlt werden. Die Berechnung der Vorgabetemperatur T2 erfolgt so mit Hilfe eines Fuzzy-Logic Systems. Dies bietet dem Anwender die Möglichkeit, sein Wissen über die notwendige Kühltemperatur in das Fuzzy-Logic-System einzubringen. Regelungstechnisch gesehen wird hier eine mehrdimensionale nichtlineare Kennlinie programmiert, die zur Führung des Regelkreises dient.

5.4 Weitere Einsatzgebiete

In dem Gebiet der Verfahrenstechnik ist nicht nur eine stabile Regelung nichtlinearer Prozesse wichtig, sondern in vielen Fällen auch die Verbindung von Expertenwissen mit regelungstechnischen Systemen. In vielen Fällen ist es wünschenswert, solches Expertenwissen in regelungstechnische Systeme einzubringen. Ein Beispiel ist die energieoptimale Führung eines Prozesses. Das hier benötigte Wissen übersteigt die Grundlagen der Regelungstechnik und beinhaltet umfangreiche prozeßspezifische Kenntnisse, die in Form unscharfer Regeln integriert werden können.

Andere Einsatzgebiete liegen in der Verknüpfung von Informationen, die aus verschiedenen Sensoren und Prozeßgrößen stammen. Diese Informationen können sowohl zu Zwecken der Identifikation, der Qualitätssicherung oder der Aufbereitung von Regelgrößen verarbeitet werden. Die Fuzzy-Logic eignet sich auf Grund ihrer Struktur auch besonders gut dazu, Regelsysteme für unscharfe technische Größen zu entwerfen. Viele technische Größen, die zur Führung von Prozessen eingesetzt werden, entsprechen einem Erfahrungswissen. Dieses Erfahrungswissen ist seiner Natur nach unscharf. Andererseits sind die benötigten Größen nicht exakt meßbar, weil die Signale entweder mit Störgrößen überlagert sind oder weil eine exakte Messung der eigentlich gewünschten Meßgröße in einer Anlage nicht möglich ist. Hier ermöglicht die Fuzzy-Logic eine entsprechende Aufbereitung von Meßdaten.

Andere Anwendungsmöglichkeiten liegen in der Identifikation. Hier ist es unter anderem interessant, klassische Identifikationsverfahren mit Hilfe der Fuzzy-Logic zu verbessern. So können klassische Identifikationsverfahren normalerweise nicht über die Qualität der Eingangsdaten entscheiden. Nimmt man Expertenwissen über eine Anlage hinzu, so liegt oft ein unscharfes Wissen darüber vor, ob ein Meßwert plausibel ist und sinnvoll für eine Identifikation benutzt werden kann. Die Aufbereitung einer Plausibilitätsgröße kann daher sehr gut mit der Fuzzy-Logic erfolgen. Die eigentliche Identifikation von statischen oder dynamischen Größen des Systems kann dann mit einem klassischen Identifikationsverfahren erfolgen.

Literatur

[1] Gariglio, D.: Fuzzy in der Praxis. Elektronik Nr 20, 63-75, 1991
[2] Ingenieurbüro Dr.-Ing. D. Schulz, Seminarskript, Reglerentwicklung mit Fuzzy Logik - Elementen, 1992

Regeln mit 'Fuzzy-Expertensystemen'

H. Knappe
- Technische Leitung -
Fa. SCHRÖDER airporttechnik
Mittenheimerstr. 74
8042 Oberschleißheim

Fuzzy-Expertensysteme

Expertensysteme werden in den letzten Jahren in der Industrie immer häufiger eingesetzt. Diese KI-Systeme sind aber bei weitem nicht ausgereift und immer noch Gegenstand der Forschung. Eines der größten Probleme ist heute noch die Modellierung von Unsicherheit in den Daten und der Wissensbasis (widersprüchliche Aussagen / Regeln). Es liegt daher nahe die Fuzzy-Set-Theorie, die den Anspruch erhebt, Unschärfe von Daten und sprachlichen Formulierungen bewältigen zu können, in Expertensystemen einzusetzen.

Expertensysteme sind im wesentlichen dadurch gekennzeichnet, daß Steueralgorithmus (Problemlösungsstrategie), Wissensbasis (Regeln) und Datenbasis (Problemdaten) vollständig voneinander getrennt sind. Nicht jedes Programm, das einen Experten unterstützt ist damit automatisch ein Expertensystem. Ein besserer Begriff ist daher 'wissensbasiertes System'. Expertensysteme sind im wesentlichen durch die in Abbildung 1 gezeigten Module gekennzeichnet :

- Inferenz-Maschine (Kontrollsystem für die Anwendung der Regeln, d.h. erweiterbares Regelsystem (Umgang mit unsicheren Daten sollte möglich sein)
- erweiterbare Datenbasis (für scharf abgegrenztes Wissensgebiet)
- Erklärungskomponente (transparente Begründung der Expertensystemantwort)
- komfortable Benutzerschnittstelle für die Ein-/ Ausgabe von Fragen / Antworten (natürlichsprachlich)
- einfache Schnittstelle zur Erweiterung der Wissenskomponente

Die Inferenz-Maschine ist für die Ableitung einer Anwort aus vorgegebenen Eingabebedingungen mit Hilfe der Wissensbasis verantwortlich. Sie ist daher für die Leistungsfähigkeit des wissensbasierten Systems von zentraler Bedeutung. Man unterscheidet zwischen deterministischen und nicht-deterministischen (üblicherweise probabilistischen) Ansätzen. Während deterministische Ansätze nur mit sicheren Bedingungen arbeiten können (binäre Logik), sind probabilistische Schlußfolgerungssysteme in der Lage, auch mit unsicheren Daten und Wissen umzugehen. Derartige Inferenz-Maschinen verwenden aufwendige Verfahren der Wahrscheinlichkeitstheorie.

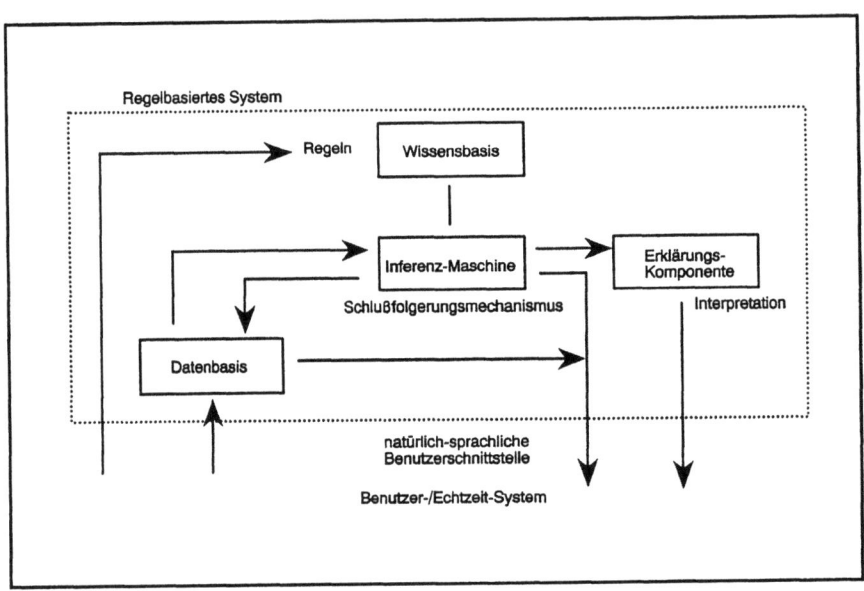

Abbildung 1: Komponenten eines Expertensystems

Bisher wurde Fuzzy-Logik (multivariable, kontinuierliche Logik) in Expertensystemen kaum verwendet. Und dies, obwohl mit dieser Logik die Erfassung und Verarbeitung unscharfer Information möglich ist. Die Ursache liegt wohl in der Tatsache, daß KI-Experten bisher an der symbolischen Informationsverarbeitung festhalten und sich gegen 'zu viel' Mathematik wehren. Durch die Publizität der Fuzzy-Logik findet aber auch hier ein Umdenken statt.

Der Einsatz von Fuzzy in Expertensystemen liegt zum einen in der Forderung nach einer natürlichsprachlichen Schnittstelle zwischen Programm und Benutzer begründet (unscharfe Aussagen/Daten). Auf der anderen Seite ist es oft nicht möglich eine widerspruchsfreie und vollständige Wissenskomponente zu erstellen. Das menschliche Wissen ist in gewisser Sicht eben unscharf.

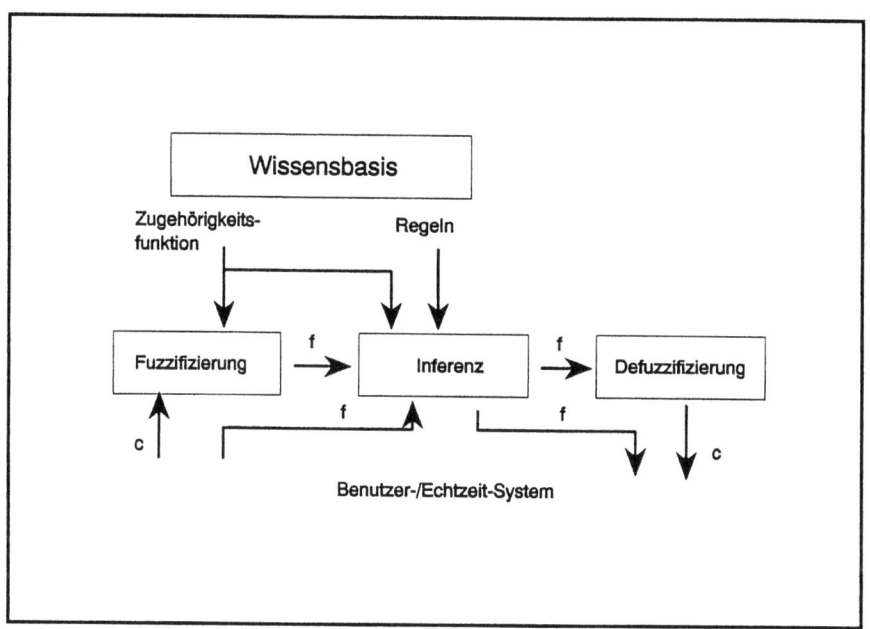

Abbildung 2: Komponenten eines Fuzzy-Systems

Abbildung 2 zeigt den Aufbau eines Fuzzy-Systems. In der Fuzzifizierungs-Komponente wird die in Zahlenwerten ('crisp' - c) gegebene Information in Wertigkeiten sprachlicher Aussagen ('fuzzy' - f) übersetzt. Dies geschieht mit sogenannten Zugehörigkeitsfunktionen, die für jede Aussage definiert werden müssen. Sie liefern einen Wert, wie gut eine Aussage über eine Variable bei einem bestimmten Zustand übereinstimmt. Die Fuzzy-Inferenz-Maschine verknüpft diese Werte mit Hilfe linguistischer Operatoren (Und/ Oder/ Mischformen). Das Inferenz-Ergebnis berücksichtigt alle Regeln der Wissenskomponente, auch wenn diese nicht widerspruchsfrei oder unvollständig sind. Allerdings kann die Schlußfolgerung noch nicht interpretiert werden. Dazu ist zuerst ein Übergang zurück in den 'crisp'-Bereich notwendig. Dies geschieht in dem Defuzzifizierungs-Schritt.

Die Aufgabenstellung einer Regelung ist der Fuzzy-Entscheidungsunterstützung (Expertensystem) sehr ähnlich. Im Prinzip muß für einen gegebenen Systemzustand eine geeignete Änderung der Stellgrößen durchgeführt werden. Dazu ist eine Interpretation und Bewertung des Zustands mit Hilfe von Regeln denkbar, besonders, wenn man an ungenaue, unscharfe Meßdaten oder widersprüchliche Systemanforderungen denkt, die verarbeitet werden müssen. Die Regelung technischer Systeme kann also auch als Aufgabe eines Expertensystems aufgefaßt werden. Anders als lineare Verfahren der klassischen Regelungstechnik, die aufbauend auf mathematischen Regler- und Systembeschreibungen eine Reglersynthese durchführen, kann mit Hilfe von Fuzzy eine völlig neue Informationsform in die Regelung einfließen. Statt der Suche nach mathematischen Modellen und der Lösung von Differentialgleichungssystemen, können verbal formulierbare Aufgabenstellungen einfach in die Regelung eingebracht werden. Der Reglerentwurf und die Inbetriebnahme wird durch die Transparenz des Fuzzy-Konzepts wesentlich beschleunigt (time-to-market). Die Berücksichtigung von dynamischen Systemeigenschaften ist bei einer Regelung sehr wichtig. Anders als bei Expertensystemen üblich, ist weniger die einzelne Entscheidung als vielmehr deren Abfolge von Bedeutung.

Das Potential von Fuzzy-Regelungen liegt zum einen in der Möglichkeit, verbale Regelstrategien in eine Regelung zu integrieren. Auf der anderen Seite sind Fuzzy-Regler nichtlinear. Dies ermöglicht eine (fast) freie Modellierung des Zustandsraums und führt bei nichtlinearen Systemen zu besserem Führungs-/ Störverhalten und höherer Robustheit (und in der Realität ist jedes System mehr oder weniger nichtlinear) gegenüber linearen Regelungsansätzen. Bei der Auslegung eines Fuzzy-Controllers müssen eine Reihe von Parametern eingestellt werden. Die Parameterauswahl ist heuristisch. Daher liegt es nahe, diese Einstellung zu automatisieren und adaptive Fuzzy-Regelungen einzusetzen. Die Forscher sehen in der Kombination von Neuronalen Netzen und Fuzzy-Logik eine Möglichkeit, die Vorteile beider Konzepte, nämlich Lernfähigkeit und verbale Regelbeschreibung, miteinander zu verbinden. Allerdings ist dieses sogenannte 'Neuro-Fuzzy' für den industriellen Einsatz noch ungeeignet, da geeignete Optimierungskriterien bisher noch fehlen.

Der Fuzzy-Reglerentwurf ist nicht unproblematisch. Die Erstellung der Regelbasis gestaltet sich bei vielen Ein-/Ausgangsgrößen schwierig, da die Abhängigkeiten aller Eingänge untereinander berücksichtigt werden müssen und dafür viele Regeln notwendig sind. Durch hierarischen Entwurf (z.B. als Kaskade) kann die Anzahl der Regeln allerdings reduziert werden, wenn keine vollständige Korrelation aller Eingangsgrößen vorliegt. Für Regelungen technischer Prozesse, für die nicht unmittelbar eine Regelstrategie angegeben werden kann, ist das Finden von Regeln schwierig und zeitintensiv. Dies gilt vor allem für Systeme, bei denen ein mathematisches Modell in Form von Differentialgleichungen vorliegt. Hier ist der Einsatz von Fuzzy-Logik nicht unbedingt sinnvoll.
Wenn die Regelbasis erfolgreich erstellt wurde, sind weitere Einschränkungen zu beachten. Die Wahl der Form und Lage (vor allem Überlappungsbereich) der Zugehörigkeitsfunktionen ist von entscheidender Bedeutung. Sind diese ungeschickt

gewählt, so werden die durch die Regeln exakt vorgegebenen Stellgrößen für einen bestimmten Systemzustand nicht mehr angenommen. In diesem Fall müssen die vom Fuzzy-Regler gelieferten Ausgangswerte entsprechend skaliert werden, so daß der gewünschte Stellbereich wieder vollständig ausgenutzt wird. Das Design wird vor allem durch die gegenseitigen Abhängigkeiten der Einstellgrößen des Fuzzy-Reglers (Regeln, Zugehörigkeitsfunktionen, Skalierung, Defuzzifizierungsmethode, linguistische Operatoren, Inferenz-Methode) erschwert. Deren Auswirkungen sind nur durch Simulation zu erfassen und noch weitgehend unbekannt.

Die Regeln bilden das Grobverhalten des Reglers aus (explizites Wissen). 'Regeln' für die Einstellung der Fuzzy-Parameter sind bisher noch nicht erforscht (Entwurf durch 'Probieren'). Die Feinmodellierung wird vor allem durch die Zugehörigkeitsfunktionen (implizites Wissen) festgelegt. Im Prinzip bilden die Regeln Stützpunkte im Zustandsraum (siehe Abbildung 6) des Reglers. Die Zugehörigkeitsfunktionen sind für die nichtlineare Interpolation dieser Stützpunkte verantwortlich. Je nach Form und Überlappungsbereich entstehen steile und falsche Bereiche oder auch Knicke. Diese normalerweise unerwünschten Eigenheiten des Phasenraums spiegeln sich später in Grenzschwingungen, Instabilität oder schlechtem Führungs-/ Störverhalten des Gesamtsystems wieder. Meist behilft man sich in solchen Fällen durch Einfügen von zusätzlichen Stützpunkten, d.h. Regeln, die eine bessere Annäherung und Kontrolle Phasenraums erreichen. Kriterien zur Überprüfung von Stabilität oder Verfahren zur Schwingungsdämpfung sind noch unbekannt. Die richtige Form und Lage der Zugehörigkeitsfunktionen zu finden, ist meist nur durch Probieren, Erfahrung und Fingerspitzengefühl möglich. Während die Regeln als explizite Wissensbasis vorliegen, d.h. unmittelbar verstanden und interpretiert werden können, ist dies für die implizit im Regler steckende Information nicht möglich. Die vielgelobte Transparenz von Fuzzy-Reglern, auch für den Laien, ist auf den zweiten Blick nicht unbedingt gegeben.

Schließlich ist also doch ein Experte für den Entwurf von Fuzzy-Reglern notwendig! Das Allheilmittel ist Fuzzy wohl doch nicht. Im Einsatz an der richtigen Stelle birgt die Fuzzy-Theorie, vor allem auch in Zusammenhang mit traditionellen Reglerverfahren (hybrid Fuzzy-Systeme), ein enormes Potential.

Fuzzy-Enteisungstechnik

Die Vorteile von Fuzzy-Control halten mittlerweile auch in der Industrie Einzug. Als Beispiel dient die Regelung des Mischverhältnisses und der Spitzmenge eines Flugzeug-Enteisers. Enteisungsfahrzeuge werden auf jedem Flughafen zur

Tragflächenenteisung von Flugzeugen eingesetzt (siehe Abbildung 3). An Frosttagen wird durch die Enteisung, die 10 bis 20 Minuten vor dem Start stattfindet, sichergestellt, daß die Aerodynamik des Flugzeugs durch Eis auf den Tragflächen nicht beeinträchtigt wird. Da das verwendete Enteisungsmittel teuer und umweltschädlich ist, sind die Betreiber angehalten, es mit heissem Wasser gemischt in möglichst geringer Konzentration einzusetzen. Dadurch reduziert sich allerdings die Wirksamkeit und Vorhaltezeit der Enteisung. Die sparsame Art zu Enteisen ist vor allem nur dann möglich, wenn das vorgewählte Mischungsverhältnis exakt eingehalten wird. Ein 14-Meter-Hubarm mit einem Arbeitskorb gibt dem Bediener die nötige Arbeitsposition um mittels eines Sprühmonitors das heiße Wasser-/Enteisungsmittelgemisch (ca. 80 $^\circ$C) auf die Tragflächen aufzubringen.

Abbildung 3: Flugzeug-Enteisungsfahrzeug
(hier im Einsatz für den Flughafen Düsseldorf)

Abbildung 4 zeigt das Funktionsprinzip des Enteisungsfahrzeugs der Firma Schröder airporttechnik mit Mischen des Wasser-/Enteisungsmittelgemischs auf dem Fahrzeug. Die notwendige Temperatur des Gemischs wird durch Erhitzen des Wassers mit einem Brenner vor dem Mischen erreicht (Enteisungsmittel wird durch Erhitzen über 80 C zerstört). Das eingesetzte Enteisungsmittel ist in seiner chemischen Zusammensetzung so empfindlich, daß es von scharfen Kanten zerstört wird. Aus diesem Grunde werden Exzenterschneckenpumpen zur Förderung eingesetzt. Diese weisen jedoch eine nichtlineare Abhängigkeit des Wirkungsgrads von Temperatur und Druck auf. Dadurch ist eine optimale Regelung der

Durchflußmenge nur mit Schwierigkeit zu erreichen. Das exakte Regeln wird außerdem durch Totzeitanteile und Nichtlinearitäten der Stell- und Meßglieder erschwert. Schließlich ist die Spritzmenge auch bei variabler Last durch Verändern der Düsenöffnung oder Auswahl unterschiedlicher Zuleitungen zur Düse für einen großen Mengenbereich von 0 bis 400 l/min auf 1 l/min exakt einzuhalten.

Die Exzenterschneckenpumpen werden hydraulisch angesteuert. Die Fördermenge hängt dabei sehr stark von der Temperatur der Pumpe bzw. Mediums ab. In der Spritzleitung wird der von der Pumpe erzeugte Druck und Durchfluß gemessen. Der induktive Durchflußmesser weist eine hohe Totzeit auf, so daß die Wirkung der Ansteuerung erst spät gemessen werden kann. Zusätzlich sind die Meßsignale von Durchfluß und Druck stark verrauscht.

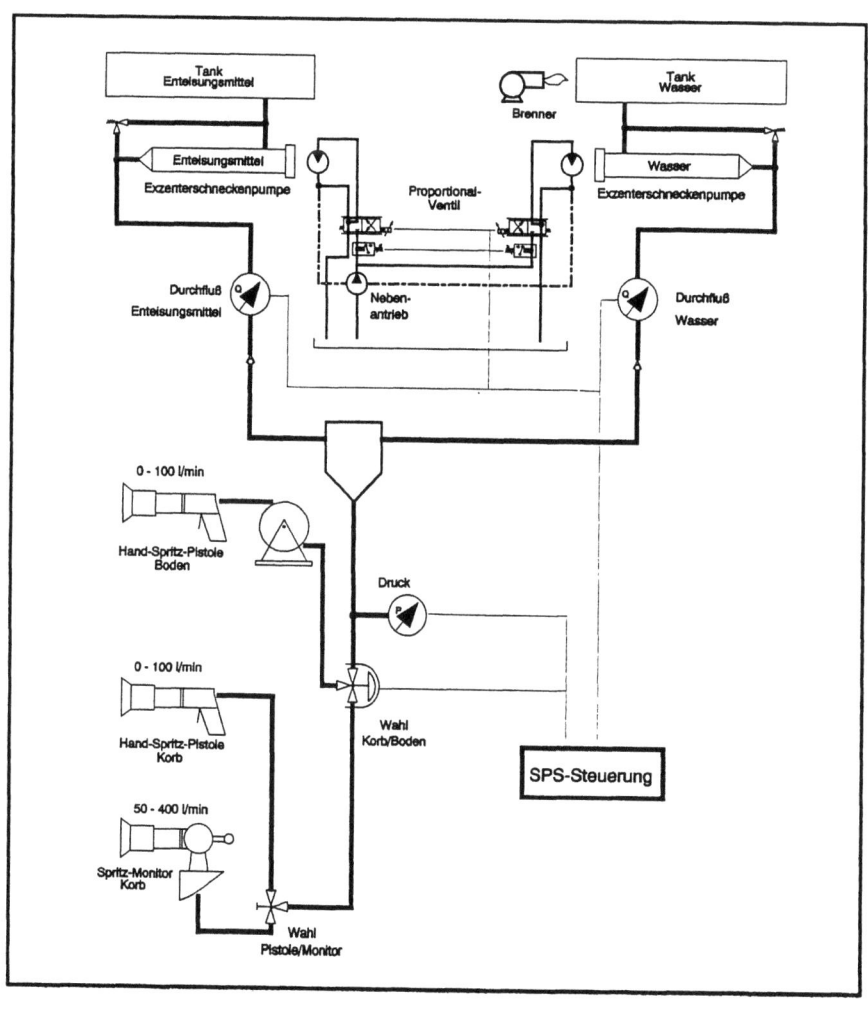

Abbildung 4: Enteiser-Funktionsprinzip

Ziel der Regelung ist es, den gewünschten Durchfluß (l/min) möglichst schnell zu erreichen und auf einen Liter genau zu halten, um das vorgewählte Mischungsverhältnis mit minimaler Abweichung zu garantieren. Da an der Spritzdüse während des Spritzens die Austrittsöffnung verstellt werden kann, ist eine Nachregelung gegebenenfalls erforderlich. Auch diese soll natürlich möglichst schnell erfolgen. Wird die Düse ganz geschlossen, sinkt der Durchfluß bei starkem Druckanstieg also auf Null, so muß die Pumpe automatisch abgeschaltet werden. Erst wenn ein Druckabfall auftritt, muß diese wieder anlaufen und geregelt den gewünschten Durchfluß einstellen oder erneut abschalten.

Dadurch ist das Anfahren der Exzenterschneckenpumpe besonders schwierig. Bevor, durch die Totzeit bedingt, ein Durchfluß gemessen werden kann, steigt der Druck stark an. Die Steuerung muß nun entscheiden, ob wieder abgeschaltet werden muß, da die Spritzdüse geschlossen ist, oder weiter geregelt werden soll. Gelöst wird dies durch Verzögerungen, die zu frühe Regelung (hohes Überschwingen) bzw. Druckabschaltung (Stöße) verhindern. Die Parameter sind abhängig von der gewünschten Spritzmenge zu wählen, da sich vor allem durch den großen Förderbereich von 0 bis 400 l/min der Druck- bzw. Durchflußverlauf verändert.

Zusätzlich zur Regelung auf Durchfluß ist eine Druckregelung vorgesehen. Die Pumpe muß beispielsweise bei einem Druck von über 14 bar in der Spritzleitung unmittelbar abgeschaltet werden. Außerdem kann eine zu hohe Spritzmenge bei zu kleiner Düsenöffnung zu unzulässigem Überdruck (größer 8 bar) führen. Tritt dies ein, muß die Pumpe statt auf Durchfluß auf einen Druck von 8 bar geregelt werden.

In Enteisern wurde bisher zur Durchflußregelung ein PI-Regler eingesetzt, dessen Parameter in Abhängigkeit von der Abweichung zwischen Soll- und Ist-Menge umgeschaltet werden (siehe Abbildung 5). In der Feinregelung ist der I-Anteil hoch gewählt, um die gewünschte Menge auf ein Liter genau stabil einzuhalten. Weicht nun der Meßwert vom Solldurchfluß zu stark ab (mehr als 3 l), so wird auf eine Grobregelung umgeschaltet, die die Pumpe schnell nachführt. Dieser Durchflußregelung ist eine PI-Druckregelung überlagert, die nach Erreichen eines Drucks von 7 bar aktiviert wird und den Druck während des Spritzens auf 10 bar begrenzt.

Steht man vor der Entwicklung eines geeigneten Reglers für ein derartiges nichtlineares System- und hat man nicht die wissenschaftlichen oder zeitlichen Möglichkeiten einer exakten Reglersynthese, so liegt die Frage nahe, ob es nicht einfachere Methoden gleicher oder sogar besserer Güte gibt.

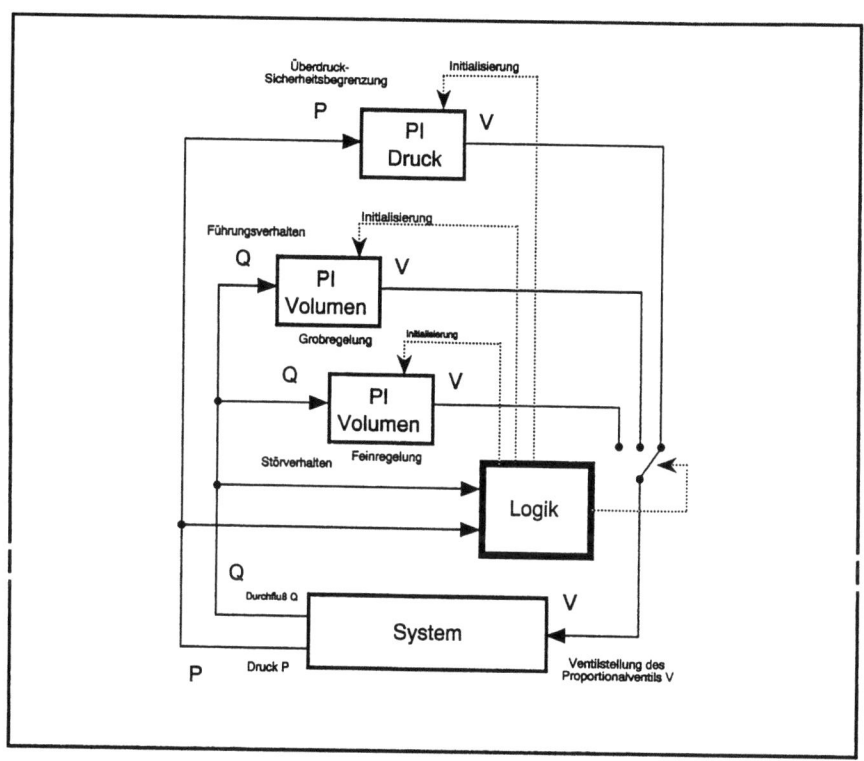

Abbildung 5: PI-Regelkonzept

Mit Fuzzy-Control war es möglich mit einer einfachen verbal formulierbaren Regelstrategie (siehe Abbildung 6) bessere Ergebnisse als mit dem PI-Regler zu erreichen. Ein wesentlicher Vorteil war die wesentlich kürzere Entwurfs- und Implementierungszeit. Vor allem konnte auf eine komplizierte Reglerumschaltung verzichtet werden, da durch die Nichtlinearität des Fuzzy-Reglers ein gutes Einschwing- und Feinregelverhalten bei hoher Robustheit (Änderung der Last, Spritzmenge und Abnutzung) mit einem Regler erreicht werden kann. Außerdem kann die Durchfluß- und Druck-Regelung gleichzeitig implementiert werden.

Typische Regeln sind zum Beispiel: Wenn der Ist-Druck wesentlich höher als der Soll-Druck ist, so muß das Proportional-Ventil, das die Exzenterschneckenpumpe ansteuert ein wenig geschlossen werden. Mit wenigen Regeln dieser Form kann die Regelstrategie erfaßt werden. Sogar die Ein-/Abschalt-Logik, die beim linearen PI-Regler getrennt realisiert werden mußte, kann in die Regelbasis eingebaut werden. Die propagierten Stell-Änderungen werden in einem Integrator aufaddiert. Der Regler ist also ein Fuzzy-P-I-Regler (nichtlineares Fuzzy-P). Abbildung 6 zeigt den Zustandsraum des Fuzzy-P-Anteils des Reglers.

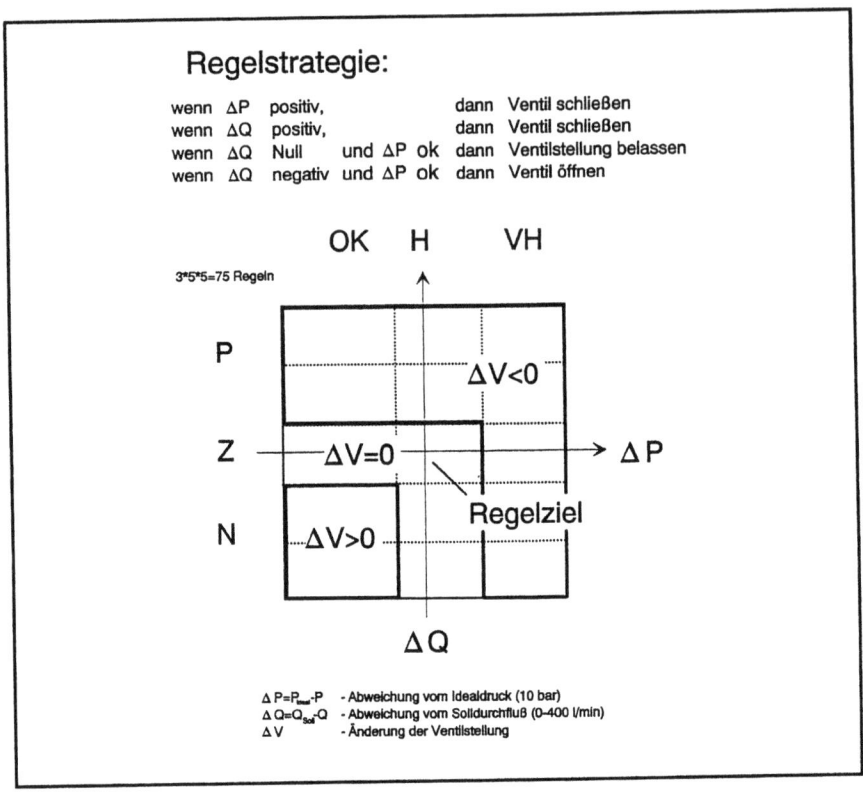

Abbildung 6: Regelstrategie und Zustandraum des Fuzzy-PI-Reglers

Abbildung 7 zeigt den gemessenen Verlauf von Druck, Durchfluß und die Ventilstellung für eine Spritzmenge von 80 l/min (Handspritz-Pistole) bei Fuzzy-I-Regelung. Die nachfolgende Tabelle zeigt eine Zusammenfassung der Nachteile des linearen PI-Regel-Konzepts und Vorteile des nichtlinearen Fuzzy-I-Reglers.

Nachteile des PI-Regelkonzepts

- zeitaufwendige Programmentwicklung
- fehleranfällige SPS-Programmierung
- PI-Anteile können nicht berechnet werden, sondern müssen durch 'Probieren' eingestellt werden (keine Optimum garantiert)
- schlechtes Führungs- und Störverhalten trotz strukturumschaltender Realisierung
- unzureichende Robustheit gegenüber Einstellbereich (Spritzmenge von 0 bis 400 l/min) und Verschleiß.

Vorteile des Fuzzy-Reglers

- extrem schnelle Entwicklungszeit
- besseres Führungs- und Störverhalten durch nichtlineare Regelung
- erhöhte Robustheit gegenüber Einstellbereich und Verschleiß
- keine getrennte Implementierung eines Reglers für Druck und Durchfluß nötig
- Logik zum Abschalten bzw. Anfahren der Pumpe kann im Regler mit implementiert werden.

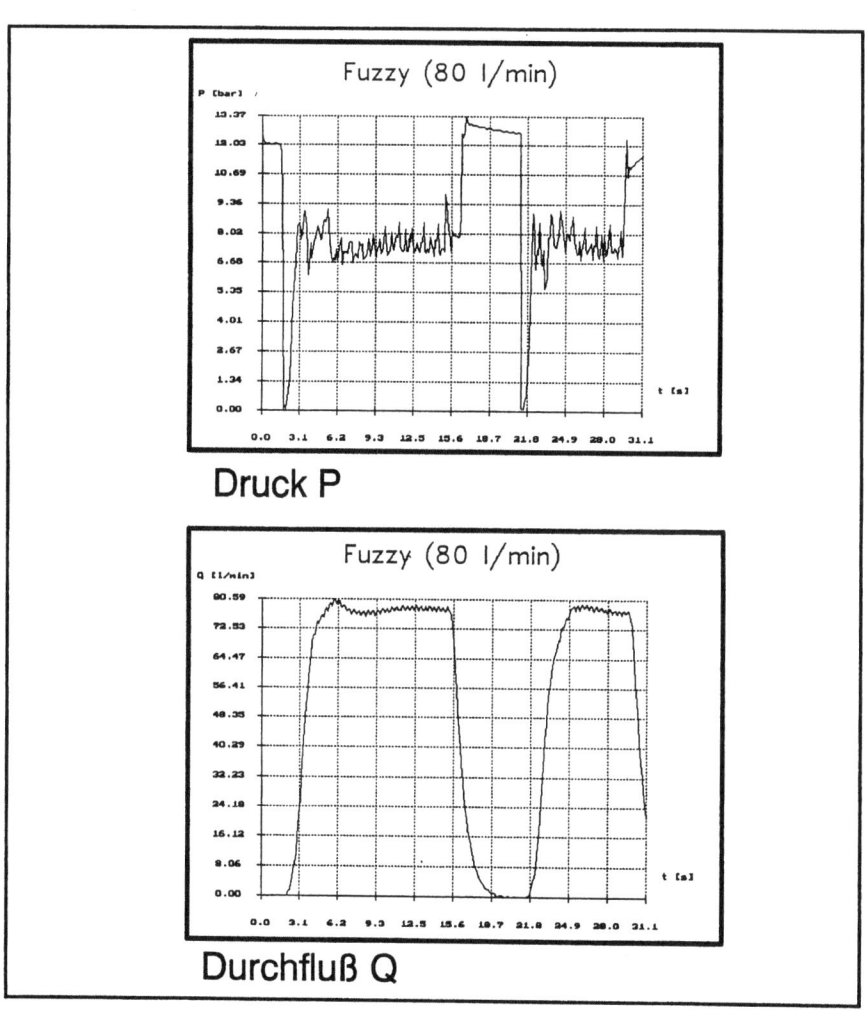

Abbildung 7: Meßdaten der Fuzzy-Regelung (Solldurchfluß 80 l/min)

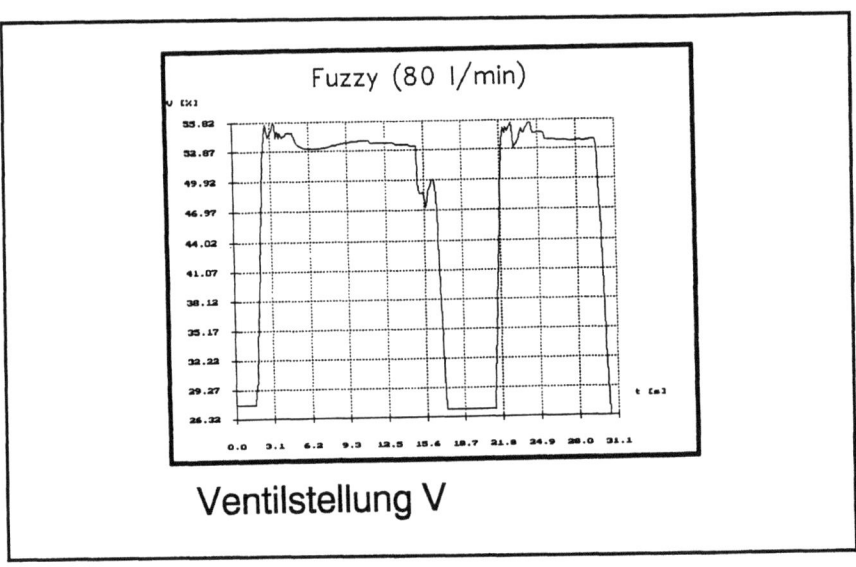

Abbildung 7: Meßdaten der Fuzzy-Regelung (Solldurchfluß 80 l/min)

Der Einsatz von Fuzzy-Konzepten für das operative
Produktionsmanagement

H.-P. Lipp   MIT GmbH Aachen
W. Ringelband HOESCH Stahl AG Dortmund

1. Das operative Produktionsmanagement/5/

Engverkettete und flexible Produktionseinrichtungen komplexer verfahrenstechnischer Prozesse sichern bei hoher Produktivität die Bereitstellung von Produkten mit kundenspezifischen Eigenschaften zu einem vom Kunden festgelegten Zeitpunkt. Die Produktionsprozeßsteuerung muß in solchen eng verkoppelten Systemen höchsten dynamischen Anforderungen genügen. Sie wird deshalb in lang-, mittel- und kurzfristigen Entscheidungsebenen ausgeführt. In den oberen Ebenen werden im Vorlauf der Produktionsphase ohne genaue Kenntnis des tatsächlichen Systemzustandes aus dispositiven und kapazitiven Betrachtungen Produktionsziele formuliert, die unter Ausnutzung der aktuellen Produkt- und Prozeßflexibilität in der prozeßnächsten Entscheidungsebene realisiert werden müssen. Die operativen Entscheidungen werden in herkömmlichen Produktionssystemen vom Schichtleiter oder vom Werkstattleiter ausgeführt und dem operativen Produktionsmanagement zugeordnet. Sie bestimmen alle produktionswirksamen Aktivitäten, die zur termingenauen und qualitätsgerechten Erfüllung des mittelfristigen Produktionsablaufes notwendig sind. Die ständig veränderten Produktionsbedingungen erfordern vorausblickende Entscheidungen, bei denen der Zeitpunkt, die Dauer und die Art der Steuerhandlungen situationsbezogen festgelegt werden. Diese operativen Steuereingriffe unterstützen nur dann optimal den Produktionsablauf, wenn sie, die augenblickliche Steuerbarkeit des Systems beachtend, zum Zeitpunkt der Störungen entwickelt werden. Der Schichtleiter benötigt dafür ein hohes Maß an technologischen Kenntnissen und einen guten Systemüberblick.

Der Einsatz von Prozeßleitsystemen für das operative Produktionsmanagement soll den Dispatcher oder Schichtleiter bei seinen unter Echtzeitbedingungen auszuführenden Entscheidungen entlasten. Die computerunterstützte Entwicklung des zu erwartenden lasten. Die computerunterstützte Entwicklung des Produktionsablaufs bereitet den Schichtleiter auf mögliche Prozeßstörungen in einem zukünftigen Betrachtungszeitraum vor, so daß er genügend große Handlungszeiträume für die Sicherstellung der Prozeßstabilität, der Produktqualität und von Terminstellungen erhält.

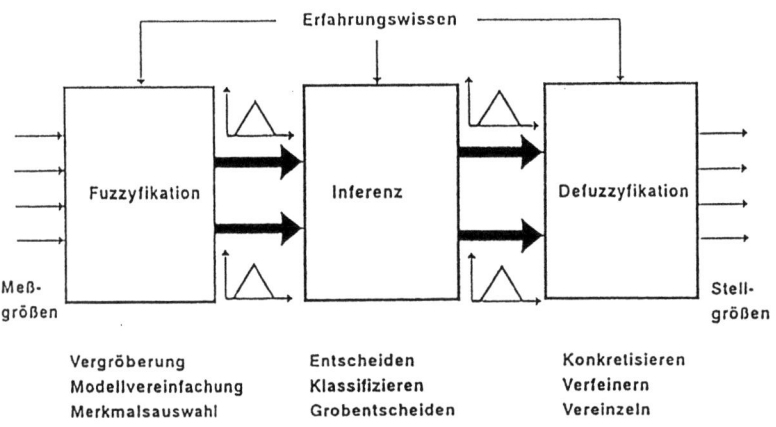

Bild 1: Entscheidungsprozeß mit Fuzzy-Methoden

Fuzzy-Konzepte in komplexen Entscheidungsprozessen
Entscheidungen , die bisher vom Menschen ausgeführt wurden, können mit Hilfe von Fuzzy-Methoden auf Rechner übertragen werden. Fuzzy-Konzepte gestatten dabei direkt im Automatisierungskonzept die Berücksichtigung von oft nur linguistisch ausgedrücktem Expertenwissen und ermöglichen gleichzeitig die effektive Ausführung von komplexen Entscheidungsprozessen. Durch die Verunschärfung oder Vergröberung von Prozeßinformationen erhält man Entscheidungsräume von geringer Dimension. Aus der Dialektik zwischen Wesentlichem und Unwesentlichem wird viel Prozeßwissen geeignet vergröbert, so daß auch in unterschiedlichsten Prozeßsituationen realitätsnahe Entscheidungen unter dem Zeitzwang des Prozeßablaufes ermittelt werden können. Das oft mit Fuzzy in Verbindung gebrachte Argument, mit möglichst wenig Wissen gute Entscheidungen treffen zu können, muß deshalb präzisiert werden.

Denn die Auswahl, welche Informationen zu vergröbern sind, setzt sehr viel Prozeßwissen voraus, um sicherzustellen, daß mit den vereinfachten Prozeßmodellen trotzdem gültige Entscheidungen entwickelt werden können. Die Zusammenfassung vieler einzelner Fakten zu einer unscharfen Menge ermöglicht dabei die Reduzierung der zu verarbeitenden Informationen, ohne daß Einzeleinheiten dabei vergessen werden. Die unterschiedliche Zugehörigkeit der Elemente in einer Fuzzy-Menge erlaubt in einer Defuzzyfizierungsphase die gezielte Vereinzelung von Einzelbezügen aus dem komprimierten Mengenzusammenhang. Durch Fuzzy-Methoden unterstützte Entscheidungsprozesse werden deshalb in folgende Abschnitte unterteilt (s.Bild 1):
- die Vergröberungs- oder Fuzzyfizierungsphase,
- die Entscheidungs- oder Inferenzphase und
- die Vereinzelungs- oder Defuzzyfizierungsphase.

In der Fuzzyfizierungsphase erfolgt die Informationsreduzierung, so daß der anschließende Problemlösungsvorgang durch die Betrachtung von unscharfen Prozeßinformationen mit der nötigen Effizienz ausgeführt werden kann. In der abschließenden Defuzzyfizierungsphase werden die unscharfen Ergebnisse des Entscheidungsprozesses vereinzelt, so daß exakte Stellhandlungen zur Steuerung des Produktionsprozesses angegeben werden können.

Die Fähigkeit, komplexe Entscheidungsprozesse durch geeignete Vergröberungen effektiv ausführen zu können, ist eine täglich zu beobachtende Fähigkeit des Menschen, die er sich in seiner langjährigen Entwicklungsgeschichte erworben hat. Um den Inhalt eines Vexier- oder anderen stark strukturierten Bildes zu erkennen, muß das menschliche Gehirn sämtliche Bildstrukturen zueinander in Beziehung bringen. Dieser Problemlösungsprozeß wird oft dadurch unterstützt, daß die zu verarbeitende Informationsmenge reduziert wird. Durch das Zusammenkneifen der Augen werden bestimmte Frequenzbereiche des einfallenden Lichtes einfach abgeschnitten, so daß unscharfe Übergänge der Bildstrukturen wahrgenommen werden. Das Gehirn hat dadurch weniger Informationen zu verarbeiten und kann den Bildinhalt schneller identifizieren.

In den nachfolgenden Betrachtungen sollen verschiedene Fuzzy-Konzepte für komplexe Entscheidungsprozesse im operativen Pro-

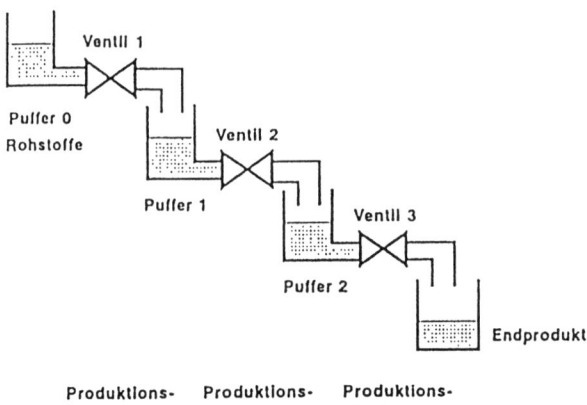

Bild 2: Beispiel für ein Produktionssystem

duktionsmanagement eingesetzt werden. Der Rechner soll mit ähnlich vergröberten Informationen versorgt werden, wie sie vom Menschen bei operativen Entscheidungen verwendet werden. Um das Verständnis dieser Betrachtungen durch spezielle prozeßbezogene Technologiekenntnisse nicht zu erschweren, wird ein vereinfachtes Produktionssystem verwendet. Dieses Beispielsystem besteht aus einer Kette von Behältern oder Puffern, deren Füllstand über Ventile beeinflußt werden kann (s. Bild 2). In diesem System besteht die Aufgabe, einen vorgegebenen Sollstoffstrom durch die Verstellung der Ventile so zu beeinflussen, daß vorgesehene Füllstandshöhen in den Puffern möglichst gut eingehalten werden. Die Pufferfüllstände kennzeichnen Arbeitsbedingungen für die angrenzenden Produktionsabschnitte und sind ein Maß für die Stabilität des gesamten Produktionsprozesses. Die Ventile präsentieren in Wechselbeziehung stehende Produktionsabschnitte, die über Puffer miteinander verkoppelt sind. Jede Ventilstellung entspricht dann einem möglichen Arbeitspunkt in Teilanlagen. Die Ein- und Ausgangsverhältnisse der Stoffströme an den Ventilen kennzeichnen notwendig einzuhaltende Rezepturen, die die Produktqualität in dem zugehörigen Produktionsabschnitt sichern.

2. Einsatz des Fuzzy-Control-Konzepts als Produktionsregler in
   komplexen Produktionssystemen

Das Fuzzy-Control-Konzept/6/ wird in Systemen eingesetzt, wo nur
minimale Strukturkenntnisse zur Lösung von Steuerungsaufgaben
verfügbar sind. Es verwendet Steuerregeln vom Typ IF ....
THEN...., die abhängig vom Istzustand Stellgliedhandlungen akti-
vieren, um die gewünschte Zustandsverbesserung zu erreichen. Die
Regeln beschreiben damit eine Zuordnungsvorschrift zwischen Systemzustand und Stellgliedhandlungen, wobei wegen fehlender Sy-
stemkenntnisse die Stellgliedaktivitäten als Repräsentanten für
den zu erreichenden Folgezustand betrachtet werden. Der Systemzustand wird häufig durch die Regelabweichungen und deren Trend
ausgedrückt, während die Stellgliedangaben entweder als inkrementale oder absolute Veränderungen angegeben werden.

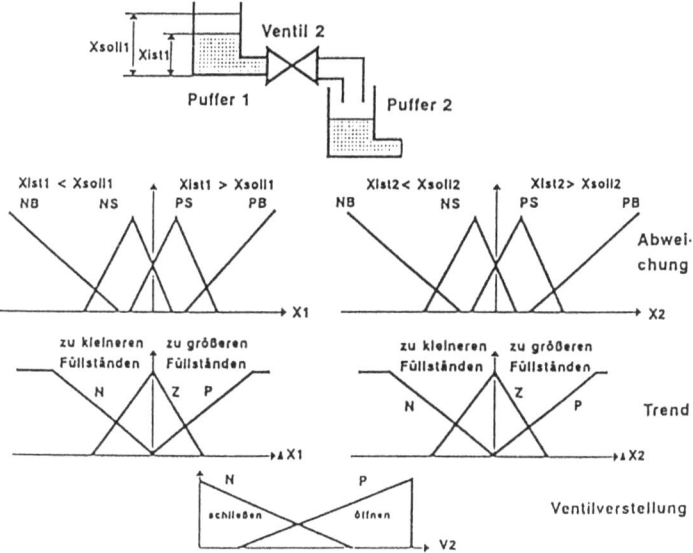

Bild 3: Bildung der Regelmenge zur Verstellung des Ventils V2

Um die Leistungsfähigkeit dieses Konzepts für operative Steuerungen komplexer Produktionssysteme zu ergründen, werden die IF-
THEN-Regeln für die abgestimmte Festlegung der Arbeitsregimes in

den einzelnen Produktionsabschnitten bzw. Ventilen unseres Beispielsystems entwickelt. Für jedes Ventil wird aus den Abweichungen und dem Trend der zugehörigen Pufferfüllstandshöhen eine Regelmenge mit den Angaben für die notwendige Ventilverstellung ermittelt. Um einer Explosion der Regelmenge entgegenzuwirken, werden dabei Teilmengen gebildet, die sich aus den getrennten Betrachtungen des Zustandes des vor und nach dem Ventil angeordneten Puffers ergeben (s.Bild 3).
Der vorher diskutierte Vergröберungseffekt wird in diesem Konzept dadurch erreicht, daß die Regelabweichung der Pufferfüllstandshöhen in die vier unscharfen Quantifizierungsstufen positiv big(PB), positiv small(PS), negativ small(NS) und negativ big(NB) eingeteilt wird. Der Trend der Regelabweichungen wird in den drei Quantifizierungsstufen negative(N), zero(Z) und positive(P) und die Stellgliedveränderung in den Mengen "Ventil öffnen" und "Ventil schließen" ausgedrückt. Durch das gleichzeitige Feuern mehrerer Regeln mit unterschiedlicher Stärke wird in der Defuzzyfizierungsphase zwischen den groben Fuzzy-Regeln interpoliert, so daß trotz des Vergröберungseffektes verfeinerte Steueraktivitäten im Prozeß wirksam werden können (s.Bild 4). Die Defuzzyfizierungsmöglichkeit der unscharfen Steuerregeln ist der Grund dafür, daß die zu erwartende Regelgüte eines Fuzzy-Controllers nicht unbedingt von der Anzahl der Regeln abhängig ist. Die aus den Quantifizierungsstufen des Zustandsraums resultierende Regelanzahl ist vielmehr eine spezielle Eigenschaft der Regelstrecke. Sie wird bestimmt durch die Art der Nachbarschaftseigenschaften des Zustandsgebietes, wie sie bei lokalen Nichtlinearitäten anzutreffen sind.
Aus der Betrachtung der beiden Teilregelmengen für das Ventil V2 wird recht bald deutlich, daß die Regeln sowohl kooperativ als auch kontrovers wirkende Stellgliedverstellungen auslösen können. Befindet sich in dem vorgelagerten Puffer des Ventils zu viel und in dem nach dem Ventil angeordneten Puffer zu wenig Stoff, so werden aus beiden Regelmengen solche Regeln aktiv, die das Öffnen des Ventils fordern. In diesem Systemzustand wird der kooperative Einfluß beider Regelmengen auf das Stellglied sichtbar. In dem Zustand, wo in beiden Puffern eine zu große Stoffmenge gespeichert ist, werden gegensätzlich wirkende Steueraktivitäten aktiviert. Aus dem Systemzustand des Puffers 1 wird dann

Ventils gefordert. Beide Steuerhandlungen führen zu einem Konflikt, der in der nachfolgenden Defuzzyfizierungsphase nur gelöst werden kann, wenn sich die Stellgliedangabe in den Regeln auf die absolute Ventilstellung und nicht wie vorgesehen auf die inkrementale Ventilverstellung bezieht.

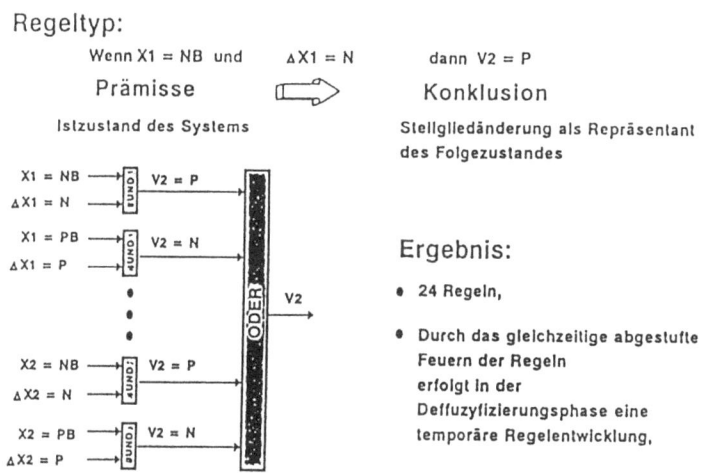

Bild 4: Regelmenge für das Ventil V2

In den Fuzzy-Control-Regeln sind die sich aus den Ventileinstellungen ergebenden Pufferfüllstandshöhen nicht enthalten. Der kooperative Einfluß eines oder mehrerer Ventile auf die Pufferzustände kann deshalb nicht oder nur mit sehr hohem Regelaufwand bei Steueraufgaben berücksichtigt werden. Dieser Mangel wird besonders dann deutlich, wenn durch das gleichzeitige Zusammenwirken mehrerer Stellhandlungen an unterschiedlichen Ventilen Störungen in Puffern in einer Vorzugsrichtung aus dem Gesamtsystem hinausgeleitet werden müssen. Das "Hineindrücken" von Störungen in das System führt oft zu erheblichen Sekundärstörungen in benachbarten Teilsystemen, die nur mit großem Aufwand kompensiert werden können.
Die Regelstruktur des Fuzzy-Control-Konzepts führt bei seiner Verwendung in komplexen Entscheidungsprozessen von Produktions-

systemen zu einem weiteren Nachteil. Neben der Kompensation von
Störungen in Puffern wird in den Teilanlagen gleichzeitig auch
ein von der Produktqualität und dem Wirkungsgrad abhängiges Arbeitsregime angestrebt. In unserem Beispielsystem bedeutet diese
Forderung, daß bei der Auswahl der Stellgliedveränderungen neben
dem Pufferzustand auch noch die Solleinstellung der Ventile beachtet werden muß. Unter Berücksichtigung der Systemstruktur
müssen deshalb mehrere Teilsysteme bzw. Ventile so koordiniert
eingesetzt werden, daß sowohl Sollpufferfüllstände als auch
Sollarbeitspunkte in allen Teilsystemen möglichst gut realisiert
werden.
Die fehlenden Strukturinformationen in den Fuzzy-Control-Regeln
sind der Grund dafür, daß kein Verschmelzen verschiedener Regeln
mit unterschiedlichem Stellgliedbezug in der Defuzzyfizierungsphase möglich ist. Die interpolierende Fähigkeit bzw. der Vereinzelungseffekt zwischen diesen Regeln ist damit nicht gegeben.
Die oft bevorzugte Eigenschaft von Fuzzy-Control, für Regelungsaufgaben nur minimale Strukturinformationen zu benötigen, muß
sich deshalb überall dort nachteilig auswirken, wo Strukturkenntnisse vorhanden sind, die bei der Steuerung des Systems zu
berücksichtigen sind. Ähnlich wie in der konventionellen Regelungstechnik, wo zwischen Einsatzgebieten eines einfachen Reglers und eines Zustandsreglers unterschieden wird, ergeben sich
die Anwendungsgrenzen in komplexen Produktionssystemen des hier
dargestellten Fuzzy-Reglers. Der Einsatz des Fuzzy-Control-Konzeptes sollte deshalb dort bevorzugt erfolgen, wo das Reglerverhalten für weniger komplexe Regelstrecken nicht funktionell
durch mathematische Gleichungssysteme beschrieben werden kann.
Dies ist immer dann der Fall, wenn die Regelmenge für Regelstrecken mit linearem und nichtlinearem Systemverhalten aus dem
linguistisch ausgedrückten Erfahrungswissen der Anlagenfahrer
mit minimalen Strukturinformationen entwickelt werden muß.

**3. Das Fuzzy-Petri-Netz-Konzept als komplexer Produktionsregler
bei der operativen Führung komplexer Produktionssysteme**

Eine noch größere Kompaktheit wird bei der Modellierung unseres
Beispielsystems erreicht, wenn die Regeln des Fuzzy-Controllers

für die Verstellung der Ventile bzw. Produktionsabschnitte jeweils zu einer komplexen Regel vergröbert und in einem Fuzzy-Petri-Netz als unscharfe Transition zusammengefaßt werden (s.Bild 5). Das Fuzzy-Petri-Netz ist ein Modellkonzept für operative Entscheidungsprozesse in komplexen Produktionssystemen, in dem Strukturkenntnisse gemeinsam mit technologischen Erfahrungen der Anlagenexperten für die Auswahl von operativen Steueraktivitäten genutzt werden.

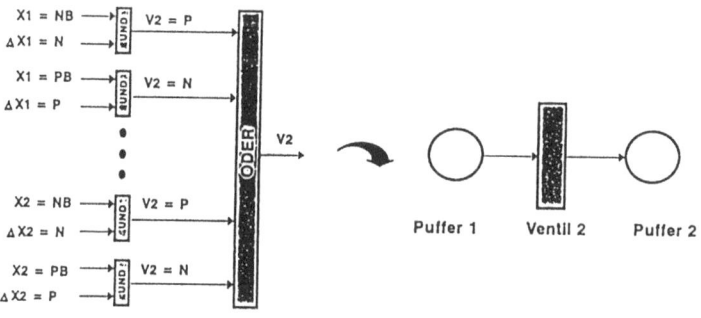

Bild 5: Regelstruktur von Fuzzy Control und Fuzzy-Petri-Netzen

Wenn der Arbeitsfortschritt in einem komplexen System durch das asynchrone Zusammenwirken mehrerer parallel wirkender Teilprozesse bestimmt wird, ist die Beschreibung der zeitlich nacheinander auszuführenden Steueroperationen auch schon durch "scharfe" Petri-Netze/1/ möglich. Die Prozeßbedingungen, wie z.B. der Füllstand der Puffer, werden als markierte Plätze, und die Steuerregeln oder Produktionsabschnitte werden als Transitionen in dem Netz abgebildet. Durch das Schalten der Transitionen entsteht zwischen den Plätzen ein Markenstrom, der zur Synchronisation des Arbeitsregimes in den einzelnen Teilanlagen benutzt wird und damit dem bisher betrachteten Stoffstrom entspricht. Diese Darstellung hat den Vorteil, daß bei unterschiedlichen

Prozeßzuständen die reale Systemstruktur und damit die kausalen Prozeßzusammenhänge der technischen Anlage weitestgehend im Modell erhalten bleiben. Das Prozeßwissen von Anlagenexperten kann damit direkt den Komponenten im Modell zugeordnet werden. Nachteilig dagegen ist jedoch das zweiwertige Verhalten der Plätze und Transitionen, wodurch Prozeßbedingungen und Steueraktionen nur in den Zuständen "zulässig/unzulässig" oder "Ein/Aus" abgebildet werden können. In realen Systemen ist dagegen durchaus ein mehr oder weniger gutes Arbeiten bei nicht ganz ideal erfüllten Prozeßbedingungen üblich. Es läßt sich in "scharfen" Petri-Netzen nicht oder nur mit einem sehr hohen Modellaufwand beschreiben. Um solche durch unvollständige Systeminformationen bedingte oder auf Erfahrungswissen begründete Handlungsfolgen dennoch im operativen Entscheidungsprozeß berücksichtigen zu können, wurde das Fuzzy-Petri-Netz-Konzept entwickelt. Mit dem Fuzzy-Petri-Netz soll ein Modellkonzept bereitgestellt werden, in dem Strukturkenntnisse - wie zulässige Arbeitsbereiche, Änderungsgeschwindigkeiten, Arbeitsbedingungen und die wechselseitige Beeinflussung von Teilprozessen auf die Qualität und die Arbeitsfähigkeit benachbarter Teilsysteme - in komplexen Entscheidungen berücksichtigt werden. Dieses Prozeßwissen resultiert aus den verfahrenstechnischen Besonderheiten und aus dem Verhalten der im Prozeß tätigen Menschen. Die Berücksichtigung dieses Wissens ermöglicht bei der operativen Führung von Produktionssystemen effiziente Entscheidungen, deren Ergebnisse mit hoher Realitätsnähe den Besonderheiten des betrachteten Prozesses entsprechen.

Das Fuzzy-Petri-Netz-Konzept /3/
Das Fuzzy-Petri-Netz wird durch die Verunschärfung der Plätze und Transitionen aus "scharfen" Petri-Netzen abgeleitet. Die Aufweichung von Bedingungskomplexen und Aktionen wird dabei mit dem Ziel vorgenommen, unvollständige Informationen und subjektives Expertenwissen bei der Modellierung von komplexen Entscheidungssituationen berücksichtigen zu können. Die scharfen Netzelemente werden unscharfen Mengen zugeordnet, wobei, da viele für bestimmte Prozeßsituationen gültige Petri-Netze zu einem Fuzzy-Petri-Netz zusammengefaßt werden, eine Modellvereinfachung erreicht wird. Die scharfen Netzkomponenten entsprechen dann mit unterschiedlicher Zugehörigkeit den Elementen im unscharfen

Netz. Für das Fuzzy-Petri-Netz gilt:

FPN = (P , T, F , S , G , M, m0)

P = {p̃}   - Menge der unscharfen Plätze
                              p̃ = { (m(p̃); µp̃(m(p̃)) }

T = {t̃}   - Menge der unscharfen Transitionen
                              t̃ = { (t;µt̃(t)) }

F : (PxT) U (TxP)-->[0,1] - Überführungsfunktion

S : (PxT)   ---> [0,1] - Startbewertung

G : (TxP)   ---> [0,1] - Zielbewertung

M : (P)   ---> N   - zulässige Markierungen

m0 ∈ M   - Anfangsmarkierung

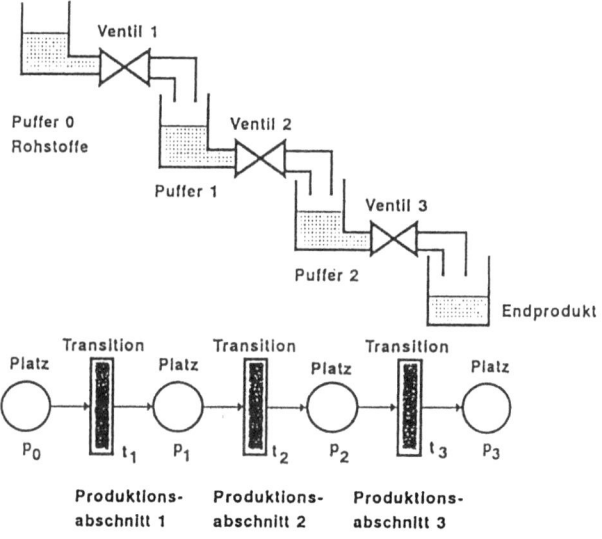

Bild 6: Fuzzy-Petri-Netz-Modell für ein Produktionssystem

Unscharfe Plätze

Um die Güte von Arbeitsbedingungen einzelner Prozeßabschnitte oder den Arbeitsfortschritt der Auftragsbearbeitung zu beschreiben, werden in dem Fuzzy-Petri-Netz unscharfe Plätze verwendet. Unscharfe Plätze stellen Fuzzy-Mengen über Platzmarkierungen dar. Die Markenanzahl in einem Platz ist ein Maß für die Erfüllung von Arbeitsbedingungen. Es werden dabei bewußt viele Marken in dem Platz zugelassen, um so den Erfüllungsgrad der Prozeßbedingungen möglichst fein unterscheiden zu können. Die unterschiedlich bewerteten Markenbelegungen dokumentieren damit einen mehr oder weniger guten Erfüllungsgrad von Prozeßbedingungen, die die Arbeitsfähigkeit der vor- und nachgelagerten Produktionsabschnitte kennzeichnen. Die unscharfe Bewertung der Platzmarkierungen durch Zugehörigkeitsfunktionen schafft eine n-wertige Stufung zwischen sehr guten und unzulässigen Arbeitsbedingungen. Die Form der Zugehörigkeitsfunktionen für die jeweiligen Plätze richtet sich nach den verfahrenstechnischen Inhalten der angrenzenden Produktionsabschnitte. Sie wird aus dem Erfahrungswissen der Anlagenfahrer abgeleitet. Breite Zugehörigkeitsfunktionen sollen eine große stabilisierende Wirkung des Platzes ausdrücken. Schmale Zugehörigkeitsfunktionen signalisieren mit einer hohen Empfindlichkeit dagegen schon geringe Abweichungen von den Sollprozeßbedingungen. Sie beschreiben Arbeitsbedingungen, die entweder aus einer vorsichtigen Fahrweise eines noch nicht eingearbeiteten Anlagenfahrers oder aus der geringen Verstellbarkeit eines komplizierten verfahrenstechnischen Prozesses resultieren. Die Plätze des Fuzzy-Petri-Netzes korrespondieren damit direkt mit den Pufferzuständen in unserem Beispielsystem.

Unscharfe Transitionen
Zur Beschreibung des Arbeitsverhaltens von Teilanlagen werden in dem Fuzzy-Petri-Netz unscharfe Transitionen verwendet. Die unscharfen Transitionen werden aus abgeschlossenen Teilsystemen oder Regeln abgeleitet. Äquivalent zum Produktstrom übertragen sie Markenströme von Eingangsplätzen in einem bestimmten Verhältnis auf Ausgangsplätze und repräsentieren die Menge von Einstellungen in einem Prozeßabschnitt. Die unscharfe Bewertung der möglichen Arbeitspunkte in einer Teilanlage gestattet die Unterscheidung von verschiedenen Arbeitsregimes. Mit möglichst hohen Zugehörigkeitswerten werden dabei Anlageneinstellungen bewertet, die durch einen hohen Wirkungsgrad, eine hohe Produktivität oder eine hohe Qualität verbunden sind. Mit sehr geringen Zugehörigkeitswerten wird dagegen die Fahrweise einer Teilanlage im Unter- oder Überlastbereich bewertet. Diese Einstellungen sollten vermieden und nur dann zur Stabilisierung des

Prozesses eingesetzt werden, wenn keine weiteren Steuermöglichkeiten verfügbar sind. Die Form der Zugehörigkeitsfunktion ist damit ein Maß für die Steuerbarkeit eines Teilsystems. Ähnlich wie bei der Platzbewertung resultiert sie aus Systemkenntnissen der Experten. Durch breite Zugehörigkeitsfunktionen wird der disponible Einsatz der Anlage eines erfahrenen Anlagenfahrers ausgedrückt. Schmale Zugehörigkeitsfunktionen weisen dagegen auf die geringe Flexibilität des verfahrenstechnischen Prozesses oder auf die vorsichtige Fahrweise eines wenig eingearbeiteten Anlagenfahrers hin.

Unscharfes Schalten

Um die gegenseitige Einflußnahme der Teilanlagen beschreiben zu können, wird jede Transitionseinstellung mit einem bestimmten Markenstrom verbunden. Der Markenstrom kennzeichnet damit auch Nebenwirkungen, die die Arbeitsfähigkeit und den Qualitätsverbund miteinander in Wechselwirkung stehender Teilsysteme beeinflussen. Er ermöglicht damit die flußabhängige Anpassung der Platzmarkierungen in Abhängigkeit von der Fahrweise der Teilanlagen. Bei nicht erfüllten Produktionsbedingungen wirkt die unscharfe Platzbewertung wie eine ereignisabhängige Triebkraft auf die Verstellung der angrenzenden unscharfen Transitionen. Der Wechsel von einer Transitionseinstellung in eine andere wird als unscharfes Schalten bezeichnet. Gegenüber einer scharfen Transition, die nur die Zustände "kein Markenstrom" und "Schalten mit einer bestimmten Markenanzahl" beinhaltet, kann bei ihr ein mit unterschiedlicher Markenanzahl n-wertiges Schalten realisiert werden. Die unscharfe Bewertung der dynamischen Schaltfähigkeit eines Teilsystems berücksichtigt dabei Expertenerfahrungen, die aus den dynamischen Eigenschaften der lokalen verfahrenstechnischen Prozesse resultieren. Sie wirkt sich hemmend auf die Triebkräfte der Plätze aus, um eine Prozeßunruhe zu vermeiden, die sich aus voreiligen Reaktionen bei Störungen ergeben. Die situationsbezogene optimale Verstellung der Teilaggregate in einem Produktionssystem wird dann aus der Wechselwirkung zwischen den Puffertriebkräften und den hemmenden Kräften der Transitionen bestimmt.

Unscharfe Startbewertung

In welcher Reihenfolge Verstellungen an Teilanlagen vorgenommen werden, um Prozeßstörungen zu beseitigen, ist von der Struktur der Gesamtanlage abhängig. In dem Netzmodell werden die aus lokalen und globalen Strukturkenntnissen abgeleiteten Informationen in der unscharfen Startbewertung $S(\bar{p}x\bar{t})$ der Kanten zwischen den Plätzen und Transitionen berücksichtigt. Die Startbewertung ermöglicht das Verschmelzen mehrerer Steuerhandlungen mit unterschiedlichem Stellgliedbezug in der Defuzzyfizierungsphase. Sie unterstützt ein schrittweises Defuzzyfizieren, so daß eine Strukturunschärfe beim Schalten unterschiedlicher Transitionen erreicht wird.

Unscharfe Zielbewertung

Die Zielbewertung $G(\bar{t}x\bar{p})$ des unscharfen Netzes dient zur Auswahl von unscharfen Schaltfolgen an den Transitionen unter Beachtung des augenblicklichen Netzzustandes. Abhängig vom Anwendungsfall soll sie Lösungen ermöglichen, die neben einem minimalen Zielabstand auch ein gutes dynamisches Lösungsverhalten garantieren.

In einem Fuzzy-Petri-Netz stellt eine Schaltfolge das wiederholte unscharfe Schalten der Transitionen dar. Sie überführt das Petri-Netz von seinem Ist- in einen Zielzustand. Die Schaltfolge der Transitionen wird von einem Problemlöser bestimmt, der den Schaltzeitpunkt und die Größe der Transitionsänderungen situationsbedingt so festlegt, daß bei möglichst guten Platzbedingungen der Zielmarkenstrom erreicht wird. Der Problemlöser ist dabei durch die unscharfen Netzbewertungen in seinem Lösungsverhalten veränderlich.

In Bild 7 wird die Arbeitsweise des Fuzzy-Petri-Netzes in unserem Beispielprozeß deutlich gemacht. Die einseitige Platzbewertung des Puffers $\bar{p}0$ drückt den Bearbeitungszustand des Produktionsauftrages aus. Die Markenbelegung des Platzes beschreibt die zu verarbeitende Rohstoffmenge. Von der Platzbewertung wird solange eine schiebende Triebkraft auf den gesamten Produktionsprozeß ausgehen, bis alle Marken in dem Platz verbraucht wurden. Die ebenfalls einseitig formulierte Zugehörigkeitsfunktion des Puffers $\bar{p}2$ wirkt dagegen bis zum Abschluß des Auftrages als ziehende Kraft auf die vorgelagerten Produktionsabschnitte. Die schmale symmetrische Zugehörigkeitsfunktion des Puffers $\bar{p}1$ signalisiert Pufferfüllstandsabweichungen mit hoher Empfindlichkeit. Sie sichert damit die kontinuierliche Arbeitsfähigkeit der angrenzenden Produktionsabschnitte $\bar{t}1$ und $\bar{t}2$. Die Arbeitsberei-

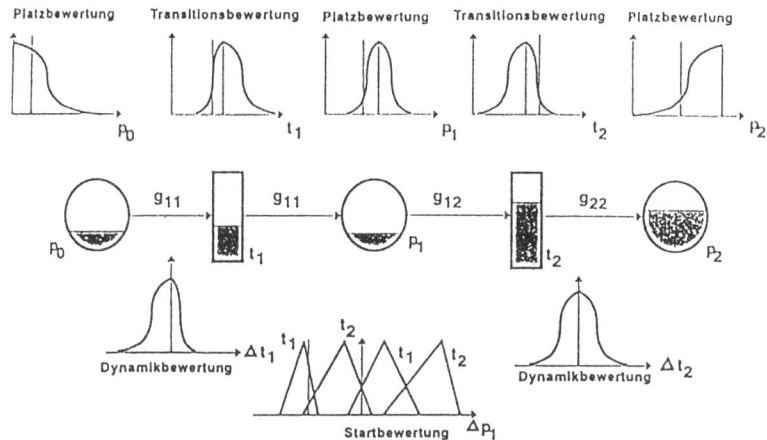

Bild 7: Unscharfe Mengen in einem Fuzzy-Petri-Netz

che der Produktionsabschnitte bzw. der Ventile wird durch die absolute und dynamische Bewertung an den Transitionen t_1 und t_2 ausgedrückt. Aus ihrer Form wird deutlich, daß der Teilprozeß t_2 oberhalb seines Sollarbeitspunktes eingeschränkt ist.

Bei der Verbesserung des Füllstandes in Puffer $\bar{p}_1$ signalisiert die zugehörige Startbewertung in Abhängigkeit von der Größe der Füllstandsabweichung, daß die Erhöhung des Markeninhaltes vorrangig durch die Verstellung des Produktionsabschnittes t_1 und dann erst durch t_2 erfolgen sollte. Diese Steuerabsicht wird durch die Transitionsbewertung von t_1 unterstützt und gleichzeitig von der Dynamikbewertung eingeschränkt. Sollte die momentane Steuerbarkeit von t_1 für die Störungsbeseitigung im Puffer $\bar{p}_1$ nicht ausreichen, dann würde die Starbewertung zusätzlich die Transition t_2 in diese Aufgabe mit einbeziehen. Dieser Vorgang wird solange wiederholt, bis keine Verbesserung im Netz mehr möglich ist.

In Bild 8 ist das zeitliche Übergangsverhalten einer aus drei Puffern und vier Ventilen bestehenden Kette abgebildet, das sich bei der Erhöhung des durchlaufenden Stoffstromes ergibt. Die

Sollhöhe möglichst konstant gehalten werden. In einem realen Produktionssystem entspricht diese Aufgabenstellung einer Produktivitätsanpassung, wie sie z.B. nach einer Reparaturphase notwendig wird. Die Pufferfüllstandshöhen und die Ventilverstellungen werden als Kurvenzug präsentiert. Sie kennzeichnen den zeitlichen Ablauf von Prozeßbedingungen mit den Arbeitspunkteinstellungen an den einzelnen Produktionsabschnitten, um den gewünschten Zustandswechsel im System zu realisieren. In Abhängigkeit von den unscharfen Bewertungen an den Puffern und Ventilen kann dabei zwischen einem aperiodischen und schwingenden Übergangsverhalten unterschieden werden.

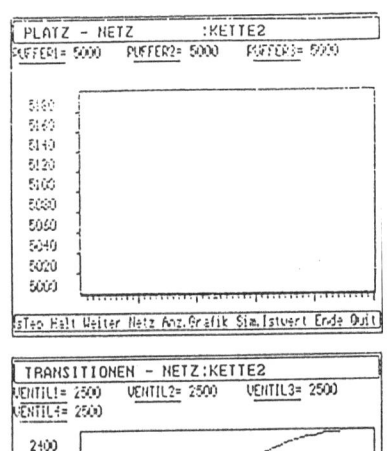
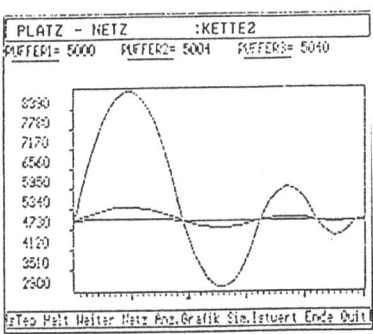
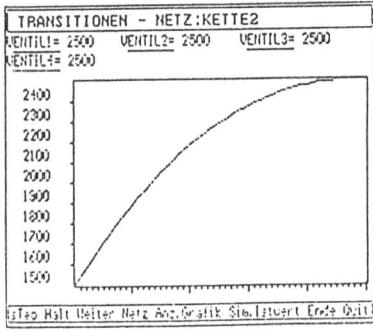
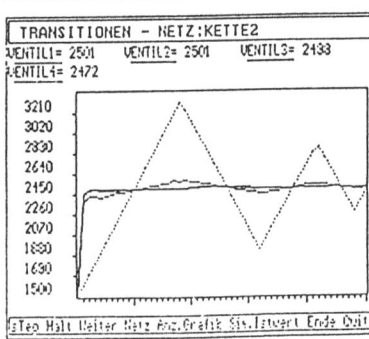

a) gleiche Steuerbarkeit der Teilsysteme

b) eingeschränkte Steuerbarkeit des Ventils 4

Bild 8: Dynamisches Übergangsverhalten bei einer Produktivitätsumstellung

Im linken Teil des Bildes wird das Übergangsverhalten für ein System dargestellt, in dem alle Prozeßabschnitte über das gleiche statische und dynamische Arbeitsvermögen verfügen. Die Pufferfüllstandshöhen sind durch eine relativ schmale Zugehörigkeitsfunktion bewertet worden, um nur kleine Produktivitätsunterschiede zwischen den Teilsystemen zuzulassen. Aus dem Kompromiß zwischen der Pufferfüllstandsbewertung und der Bewertung des Arbeitsvermögens an den Ventilen ergibt sich die Verstellung am ersten Ventil der Kette, die mit zunehmender Zielnähe immer kleiner werdend, einen aperiodischen Übergang im gesamten System erzeugt. Die Verstellung der übrigen Prozeßabschnitte resultiert aus der Kompensation der sich ergebenden Pufferfüllstandshöhen. Die Stoffstromveränderungen werden aufgrund der schmalen Pufferbewertungen wie in einem sehr starr verkoppelten System direkt an den jeweils nachfolgenden Prozeßabschnitt weitergereicht. Die Füllstände der Puffer bleiben deshalb während des gesamten Übergangsprozesses konstant.

In der rechten Bildhälfte ist das Übergangsverhalten für die gleiche Produktivitätsänderung in einem System mit ähnlichen Eigenschaften dargestellt worden, dessen viertes Ventil jedoch sehr stark in seiner dynamischen Steuerbarkeit eingeschränkt ist. Die ersten drei Ventile bzw. Produktionsabschnitte versuchen, aufgabengemäß die Produktionserhöhung schnellstens zu verwirklichen. Das vierte Ventil kann wegen seiner begrenzten Steuerbarkeit jedoch diesen Produktivitätszuwachs nicht realisieren. In dem Puffer 3 baut sich eine entsprechend große Füllstandshöhe auf, die erst zeitlich verspätet abgebaut werden kann. Die zeitlich begrenzte Durchlaßfähigkeit des vierten Prozeßabschnittes führt auch zu Füllstandsschwankungen im Puffer 2 und ebenfalls zu Produktivitätsauswirkungen in den Prozeßabschnitten 2 und 3. Es kommt zu sehr großen Produktionsschwankungen, die, wenn die dynamische Verstellbarkeit der übrigen Ventile gegenüber dem ersten Beispiel nicht verbessert worden wäre, zu erheblichen Zeitverzögerungen bei der Ausführung der gewünschten Produktivitätsumstellung führen würde.

Eine solche Vorwärtssimulation des Prozeßablaufes kann in einem Prozeßleitsystem als Erklärungskomponente für operative Produktionsentscheidungen verwendet werden, um den Schichtleiter auf mögliche Störungen oder auf die Realisierbarkeit seiner Zielstellungen vorzubereiten. Der Schichtleiter erhält dadurch einen genügend großen Handlungszeitraum für korrigierende Maßnahmen, die zur Sicherstellung der Prozeßstabilität notwendig sind.

Bei off-line Entscheidungen kann ein solches Fuzzy-Petri-Netz-Modell auch für Konfigurierungsaufgaben von Produktionssystemen verwendet werden, um durch die dynamische Paßfähigkeit der Teilprozesse einen qualitätsgerechten Verbund der Teilanlagen sicherzustellen.

Zeitbewertete unscharfe Transitionen/3,4/

Um Schaltzeiten, Totzeiten, Verzögerungszeiten u.a. bei operativen Produktionsentscheidungen berücksichtigen zu können, wird das Fuzzy-Petri-Netz zum zeitbewerteten Fuzzy-Petri-Netz erweitert. Der Schaltvorgang der Transitionen wird dann aus einer Initialisierungs-, Ausführungs- und Beendigungsphase zusammengesetzt. Die Unschärfe des Fuzzy-Petri-Netzes ermöglicht dabei die Beschreibung von Regeln oder Prozeßabschnitten, deren durch Bearbeitungs- oder Wartezeiten gekennzeichnete Prozeßabläufe zeitvariabel begrenzt werden, und bei denen in Abhängigkeit von der tatsächlich benötigten Prozeßzeit Folgebedingungen so gesetzt werden, daß mit unterschiedlichen Reaktionen der Vorgang fortgesetzt werden kann. Damit besteht mit dem zeitbewerteten Petri-Netz eine Modellstruktur, in der das bedingungs- und zeitabhängige Verhalten eines Systems nebeneinander in einem Modell beschrieben werden kann. Der Einfluß von zeitvariabel begrenzten Aktionen auf globale Systemressourcen kann dadurch in dem operativen Entscheidungsprozeß berücksichtigt werden.

4. Zusammenfassung

Fuzzy-Konzepte unterstützen das operative Produktionsmanagement in komplexen Produktionssystemen durch die Einbeziehung von Expertenerfahrungen und durch einen Vergröberungseffekt der zu verarbeitenden Informationen. In ähnlicher Weise wie der Mensch wird der Computer durch sie befähigt, komplexe Entscheidungsprozesse effektiv auszuführen. In Abhängigkeit von der Komplexität des Prozesses und den verfügbaren Informationen können das Fuzzy-Control- und das Fuzzy-Petri-Netz-Konzept in Automatisierungskonzepten eingesetzt werden.

Das Fuzzy-Control-Konzept sollte dabei für Regelstrecken eingesetzt werden, über deren dynamisches Verhalten nur sehr wenig Strukturinformationen verfügbar sind und deren Steuerregeln, wegen der fehlenden mathematischen Systemgleichungen, aus dem linguistisch ausgerückten Erfahrungswissen abgeleitet werden müssen.

Das Fuzzy-Petri-Netz-Konzept sollte dagegen bei stark strukturierten Prozessen eingesetzt werden, bei denen Strukturinformationen zwischen einzelnen Teilanlagen bewußt für eine effektive Entscheidungsfindung verwendet werden sollen. Das Fuzzy-Petri-Netz-Konzept entspricht damit dem Zustandsregler in der herkömmlichen Regelungstechnik. Der zu steuernde Prozeß wird nicht wie bei Fuzzy-Controllern als black-box-System, sondern als ein Verbund von Teilsystemen mit ihren gegenseitigen Wechselwirkungen betrachtet. In dem Fuzzy-Petri-Netz werden die Strukturinformationen in unscharfen Mengen ausgedrückt. Es wird dadurch eine sehr hohe Modellkompaktheit erreicht, so daß sich dieses Modellkonzept als Wissensrepräsentation in komplexen Entscheidungsprozessen von Prozeßleitsystemen eignet.

5. Literatur

/1/ König, R; Quäck,L.
Petri-Netze in der Steuerung
Verlag technik Berlin 1988

/2/ Lipp, H.-P.
Anwendung eines Fuzzy-Petri-Netzes zur Beschreibung von Koordinierungssteuerungen in komplexen Produktionssystemen
Wiss. Z.d.TH Chemnitz 25(1982) H.5, 633-639

/3/ Lipp, H.-P.
Ein Konzept eines unscharfen Petri-Netzes als Grundlage für operative Entscheidungsprozesse in komplexen Produktionssystemen
Diss. B, TU Chemnitz, 1989

/4/ Lipp, H.-P.
Flexible Fertigungsprozesse nach stückflußabhängigen und zeitabhängigen Kriterien steuern
ZWF CIM 1990, 12, S. 652-655

/5/ Suzaki; K.
Modernes Management im Produtkionsbetrieb
Carl Hanser Verlag München Wien 1989

/6/ Zimmermann; H.-J.
Fuzzy Set Theory and its Applications
Kluwer Academic Publishers
Boston,Dordrecht,London 1991

Fertigungsleittechnik mit Fuzzy-Logic

Ulrich Schmidt

15. Juli 1992

SDZ-GmbH, Emil-Figge-Str. 76,
4600 Dortmund 50, Tel. 0231/9742-228

1 Problemstellung

1.1 Reihenfolgeplanungsverfahren

In einer arbeitsteiligen Wirtschaft mit immer weiter zurückgehenden Arbeitszeiten der Beschäftigten kommt der Optimierung von Produktionsprozessen eine besonders bedeutsame Stellung zu. In Deutschland trifft dies vor allem auf technologisch hochwertige Produkte zu, die mit einem hohen Bearbeitungsaufwand auf teuren Maschinen produziert werden. Im letzten Jahrzehnt hat sich der Wettbewerb durch zusätzliche Anforderungen wie z.B. *Just-in-time-Lieferungen* erschwert. Hinzu kommen geänderte Philosophien im Bereich der zu optimierenden Zielgrößen einer Fertigung, wie Bestände, Durchlaufzeiten usw. Das klassische Werkstattbelegungsproblem läßt sich unabhängig davon, ob es sich um eine *flow-shop-scheduling* oder *job-shop-scheduling* Aufgabenstellung handelt, nicht ohne EDV-Unterstützung befriedigend behandeln. Von dem Zeitpunkt an, ab dem Computer durch Hochsprachen für allgemeine Problemstellungen zugänglich geworden sind, hat es zahlreiche Ansätze gegeben, Auftragsreihenfolgen automatisch nach vorgebbaren Zielkriterien zu optimieren. Das dieses Thema immer noch auf vielen Fachtagungen diskutiert wird, weist auf den nicht immer zufriedenstellenden Erfolg dieser Versuche hin. Die folgende Aufstellung umfaßt einen Teil von EDV-gestützten Verfahren für diese Problemstellung, von denen einige später im Detail diskuiert werden:

- Lineare Optimierung
- Ganzzahlige Optimierung
- kombinatorische Analyse
- Branch-and-Bound-Prinzip
- Enumerationsverfahren
- Warteschlangenverfahren

- Evolutionsstrategische Ansätze
- Manufacturing Resource Planning (MRP II)
- Fortschrittszahlenkonzept
- Kanban
- Belastunsgsorientierte Fertigungssteuerung
- OPT-System (Optimized Production Technology)
- Simulationsbasierte Systeme
- Feinsteuerung mit grafischem Leitstand
- Wissensbasierte Ansätze

1.2 Bewertung von Verfahren zur Auftragsreihenfolgeoptimierung

Die o.a. Ansätze haben gemeinsam, daß sie auf durch den Menschen vorgebbaren Kennzahlen bzw. Zielwerten beruhen. Die verbale Beschreibung und das menschliche Verständnis dieser Werte kann von einem Computerprogramm nicht ohne weiteres nachvollzogen werden. Besonders qualitative Begriffe wie *geringer Bestand, hohe Auslastung* usw. können von einem Computer mit den herkömmlichen Ansätzen nicht bearbeitet werden. Zu einem Problem wird dies vor allem dann, wenn Zielgrößen in wechselseitigen Beziehungen wie *Konkurrenz, Behinderung* oder *Kooperation* zueinander stehen. Die Ziele *geringer Bestand* und *geringe Durchlaufzeit* beispielsweise können unter verschiedenen Umständen miteinander konkurrieren oder kooperieren.

Der Großteil der momentan verwendeten kommerziellen Leitstandsysteme stellt im wesentlichen eine computergestützte Visualisierung und Aufbereitung der Daten von Produktionsplanungs- und Steuerungssystemen (PPS) und Betriebsdatenerfassungssystemen (BDE) für den Anwender dar. Der Aspekt der automatischen Optimierung einer Auftragsreihenfolge unter Verwendung von empirisch gewonnenen Erfahrungen über Systemeigenschaften wird nur in geringen Maßen berücksichtigt, so daß der Disponent nach wie vor manuell eine Auftragsreihenfolge ermittelt und das System ihn nur in der Datenverwaltung unterstützt (vgl. auch Marktübersicht in [8]). Unter automatischer Auftragsreihenfolgeplanung wird bei den marktüblichen Systemen in der Regel eine Terminberechnung (Vorwärts- bzw. Rückwärtsterminierung) oder eine Prioritätenbetrachtung verstanden. Optimierungen unter Berücksichtigung von globalen oder lokalen Zielkriterien finden nicht statt, d.h. fertigungsspezifische Gegebenheiten, die eventuell sogar variabel sind, werden nicht berücksichtigt.

Einige der kommerziell angebotenen Leitstände basieren auf den oben aufgeführten Algorithmen. Nachdem schon auf die Anwendungsprobleme von Algorithmen im allgemeinen eingegangen wurde, folgt eine kurze Darstellung spezifischer Probleme.

Enumerationsverfahren und Verfahren auf Basis von **Evolutionsstrategien** sind im praktischen Einsatz an dem hohen Rechenzeitverbrauch gescheitert. Viele der o.a.

heuristischen Verfahren benötigen für ein effektives Funktionieren spezielle Randbedingungen.

Der **Kanban-Steuerung** liegt das Prinzip sich selbst regelnder Kreise zugrunde, in die das Produktionssystem untergliedert wird. Als Regelgröße wird im wesentlichen der Materialbestand in einem Teilkreis zugrundegelegt. Ist in einem Teilkreis ein vorgebbarer Mindestbestand unterschritten, dann wird eine Nachricht abgeschickt (Kanban), die eine definierte Nachschubmenge anfordert. Wesentliche Voraussetzungen für ein erfolgreiches Anwenden der Kanban-Steuerung sind [1] [7]:

- Harmonisierung des Produktionsprogrammes mit dem Ziel möglichst gleichmäßiger kleiner Arbeitsinhalte pro Los und ablauforientierter Betriebsmittelaufstellung mit möglichst gleichem Arbeitsrhythmus im gesamten Produktionsbereich.

- Hohe Verfügbarkeit und geringe Umrüstzeiten der Betriebseinrichtungen.

Diese Voraussetzungen sind in den meisten flexiblen Fertigungseinrichtungen, die in der Regel ein stark schwankendes Auftragsspektrum bearbeiten, nicht gegeben, so daß ein erfolgreicher Einsatz der Kanban-Steuerung wenig sinnvoll ist.

Die Fertigungssteuerung mit **Fortschrittszahlen** hat ihren Ursprung in der Automobilindustrie und basiert darauf, daß in jeder Produktionszwischenstufe die Anzahl der dort eingehenden Teile gezählt wird. Meßgröße ist hier die Stückzahl. Voraussetzung ist, daß alle Teile den gleichen Arbeitsstundeninhalt haben, der Fertigungsablauf über einen längeren Zeitraum unverändert bleibt und die Kontrollstellen zum Ermitteln der Fortschrittszahlen fest plaziert sind. Dies trifft vor allem dann zu, wenn die Werkstücke auf einem Träger in einem Flußsystem befördert werden. Für eine Fertigungssteuerung, die in einer werkstattorientierten Produktion zum Einsatz kommen soll, ist dieses Verfahren daher nur bedingt geeignet.

Mit dem Begriff **OPT-System** (Optimized Production Technology) wird ein Software-Produkt bezeichnet, das in Israel entwickelt worden ist. Grundlage ist eine gegenüber der klassischen Vorgehensweise geänderte Betrachtung des Produktionsablaufes. Es wird bei der Systembetrachtung im wesentlichen von Kapazitätsengpässen ausgegangen. Der zentrale Nachteil dieses Verfahrens besteht darin, daß dem Disponenten keine Steuerungsparameter zur Verfügung stehen, mit denen er bestimmten Zielen ein vorrangiges Gewicht verleihen kann. Ferner kann die Entscheidungsfindung, d.h. die Art und Weise, wie das System zu dem optimalen Produktionsplan gelangt, nicht nachvollzogen werden. Hinzu kommt, daß es generell problematisch ist, nur von einer optimalen Auftragsreihenfolge auszugehen.

Warteschlangenmodelle haben in vielen Bereichen als Instrument der Systembeschreibung und der Simulationstechnik Verbreitung gefunden. In der Fertigungssteuerung ist ihr Einsatz bislang daran gescheitert, daß der eingeschwungene Systemzustand als eine Grundbedingung der Warteschlangentheorie in der Praxis werkstattorientierter Systeme nicht erreicht wird. Hinzu kommen als weitere, nicht erfüllbare Bedingungen der Warteschlangentheorie, daß die Ankunftsereignisse voneinander unabhängig und zufallsorientiert erfolgen sollen und keine Prioritätensetzungen bei der Auswahl einzelner Aufträge aus der Warteschlange möglich sind. Anders als einfache Warteschlangenmodelle arbeiten **simulationsgestütze Systeme**, die zwar auch auf lokalen

Auftragswarteschlangen vor einzelnen Arbeitsplätzen basieren, aber keine eingeschwungenen Systemzustand für die Reihenfolgeoptimierung benötigen (siehe auch Abschnitt Fertigungssteuerung mit Simulation).

Das Verfahren der **Feinsteuerung mit graphischem Leitstand** stellt im wesentlichen eine Aufbereitung der Netzplan- oder Balkendiagrammtechnik für Rechner mit Grafikbildschirm dar. Auf diesem Bildschirm werden die relevanten Daten für die Produktion z.b. als Balken, die den Maschinenkapazitätsbedarf angeben, graphisch dargestellt. Der Disponent hat dann die Möglichkeit, die Auftragsreihenfolge festzulegen. Der Rechner unterstützt die Optimierung der Auftragsreihenfolge nicht, so daß sie vom Disponenten iterativ ermittelt werden muß. Dies ist entsprechend langwierig und umständlich, so daß auf Störungen oder Änderungen im Betriebsablauf nicht schnell reagiert werden kann.

Die **belastungsorientierte Fertigungssteuerung** wird als Einlastungsprinzip in vielen PPS angeboten. Sie wird z.b. als nachgeschaltete Stufe einer Grobplanung nach dem MRP-II Verfahren eingesetzt. Die belastungsorientierte Fertigungssteuerung bevorzugt aufgrund ihrer Struktur eine hohe Auslastung der Fertigungsmittel und ist somit hinsichtlich dem Verfgolgen anderer Optimierungskriterien unflexibel. Es kommt hinzu, daß die Aussagen der belastungsorientierten Fertigungssteuerung über die Durchlaufzeit eines Auftrages oder Loses um so unsicherer werden, je unregelmäßiger der Zugang bzw. Abgang an den einzelnen Arbeitsplätzen ist. Verbleibende Entscheidungsfreiräume des Bedienungspersonals werden im Allgemeinen durch Anwenden der *First-Come-First-Serve-Regel* abgedeckt.

Die im Rahmen der **Fertigungssteuerung mit Simulation** bekannten Verfahren basieren auf lokalen Warteschlangen vor den einzelnen Fertigungseinrichtungen [6]. Entsprechend vorgebbaren Zielkriterien wird für jeden lokalen Bereich eine optimale Auftragsreihenfolge ermittelt und im Simulator abgearbeitet. Daraus ergeben sich die Auftragswarteschlangen für Nachfolgeoperationen. Aus diesem Auftragspool wird bei der Freigabe der Fertigungseinrichtung nach der Abarbeitung eines Loses wieder der nächste Auftrag entsprechend den lokalen Zielwerten ausgewählt usw.. Das Protokoll aller Auftragsreihenfolgen ergibt dann das Fertigungsprogramm. Nachteilig ist dieses Verfahren, weil keine globalen Zielkriterien berücksichtigt werden können und die lokalen Zielwerte bei den zur Zeit implementierten Lösungen für alle Bereiche gleich sind. In der Praxis kann als Folge davon die Termintreue der Aufträge nur mit einer erhöhten Maschinenkapazität erreicht werden.

1.3 Struktur der Fertigungssteuerung

Der dem in diesem Artikel vorgestellten Optimierungsansatz zugrunde liegende strukturelle Aufbau einer Fertigung ist in Abbildung 1 dargestellt.

Ein System für die gesamte Produktionsplanung eines Betriebes (PPS) erstellt eine Grobplanung für alle Betriebsbereiche über einen längeren Zeitraum (z. B. einen Monat oder wöchentlich). In der Anfangsphase der PPS ging man davon aus, daß sich mit dieser Methode der zentralen Planung bereits ein optimaler Fertigungsablauf erreichen läßt. In der betrieblichen Praxis stellte man jedoch fest, daß bereits nach kurzer Produktionszeit eine geplante Auftragsreihenfolge nicht mehr eingehalten werden konnte. Auf Grund

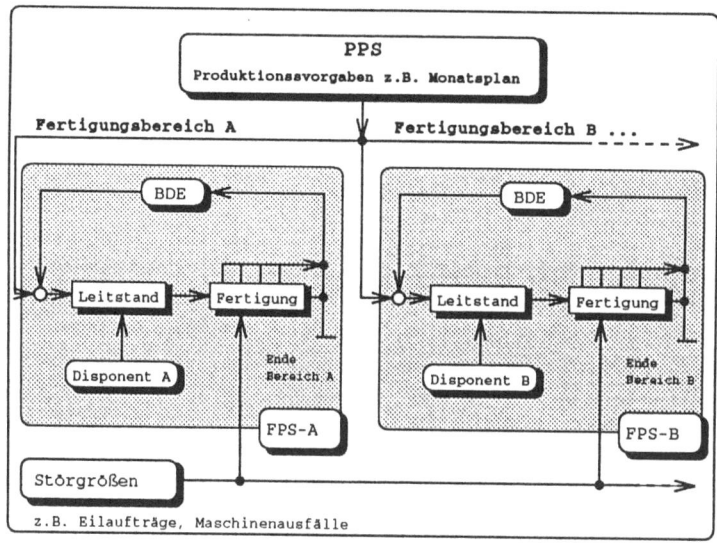

Abbildung 1: Struktur der Fertigungssteuerung

diverser Störgrößen, wie z.B. Eilaufträge, Maschinenausfälle usw., weicht die Realität meist schnell von den langfristigen PPS-Vorgaben ab, so daß sich die Notwendigkeit eines schnell, effizient und dezentral agierenden Instrumentes ergibt. In Abbildung 1 ist dieses Instrument mit dem Begriff Leitstand versehen. Dieses Konzept wurde durch den Trend zu autarken Fertigungsinseln verstärkt.

Für den hier beschriebenen Ansatz wird von der Analogie zu einem Regelkreis ausgegangen. Das PPS gibt für einen längeren Zeitraum die Sollwerte für die einzelnen Produktionsbereiche (A und B) vor (Anmerkung: Aus Gründen der Übersichtlichkeit sind in dem Beispiel die beiden Bereiche entkoppelt.). Jedem Fertigungsbereich ist in der Regel ein Mitarbeiter zugeordnet, der die Verantwortung für die Einhaltung der PPS Vorgaben trägt (z. B. Meister oder Betriebsingenieur) und der hier als Disponent bezeichnet wird. Der Disponent hat für seinen Teilbereich die Aufgabe Störgrößen auszuregeln, um die PPS-Werte zu erfüllen. Zur Unterstützung werden ihm dafür über die Betriebsdatenerfassung (BDE) Informationen über jede Produktionseinrichtung und somit auch über die Zwischenzustände der Produktion einzelner Produkte an einen Fertigungsleitstand geliefert. Leitstand, Fertigung, BDE und Disponent zusammen bilden ein Feinplanungssystem (FPS), das in Abbildung 1 grau unterlegt dargestellt ist.

Den oben skizzierten Zusammenhängen ist zu entnehmen, daß ein zentral arbeitendes PPS Schwierigkeiten hat, aktuelle bereichsspezifische Gegebenheiten in die Disposition einfließen zu lassen, zumal eine Gleichbehandlung der Bereiche A und B durch das PPS in der Regel zu keinem optimalen Ergebnis führt. Zur Zeit werden in der überwiegenden Zahl der Produktionsstätten, die auftragsorientierte Kleinserien und Einzelfertigung betreiben die Auftragsreihenfolgen manuell oder mit einem zentralem

Produktionsplanungs- und steuerungssystemen (PPS) ermittelt. Aufgrund des Zeitverbrauchs für eine Auftragsdisposition ist es unter diesen Bedingungen jedoch kaum möglich, auf plötzliche Einwirkungen oder Störungen im Produktionsablauf zu reagieren oder Besonderheiten eines Fertigungsbereiches zu berücksichtigen.

1.4 Expertensysteme

Der Mißerfolg bei der Verwendung der bisher dargestellten algorithmischen Ansätze führte zu Versuchen, die Problemstellung mit Hilfe von Expertensystemen zu lösen. Auf der Grundlage von Regeln, die das Erfahrungswissen von Experten ausdrücken sollen, versuchen diese Ansätze, Expertenwissen automatisiert auszuwerten und nutzbar zu machen.

Ansätze in den USA und in Deutschland versuchen, das Problem der automatisierten Fertigungssteuerung mit Expertensystemen zu lösen, denen das empirische Wissen über einen Fertigungsbereich als Basis zugrundegelegt wird. Während der Definitionsphase wird die Unschärfe der menschlichen Auffassung eines Problems nicht berücksichtigt. Aufgrund aktueller Ereignisse verändert sich die Entscheidungsgrundlage für die Planung der Fertigung ständig. In der Regel muß der Disponent aufgrund seiner Erfahrung Ideen zur Problemlösung entwickeln, die mit hoher Wahrscheinlichkeit noch nicht in dem Expertensystem verankert sind. In diesem Fall muß das Expertensystem aktualisiert oder die Disposition manuell durchgeführt werden.

Die Auswertung des Erfahrungswissens geschieht durch einen Inferenzmechanismus, der überprüft, inwiefern für eine aktuell vorliegende Entscheidungssituation ein als Regel erfaßtes Entscheidungsmuster in der Wissensbasis vorliegt. Liegt ein derartiges Entscheidungsmuster nicht vor, dann ist die Wissensbasis nicht vollständig, und es kann keine Entscheidung gefunden werden. Soll eine die Mehrzahl der Entscheidungssituationen überdeckende Wissensbasis aufgebaut werden, so ist die Gesamtheit der Entscheidungsmuster in ihrer Vielfalt an Kombinationen der Produktionsparameter und an möglichen Entscheidungen in Form von Regeln zu erfassen.

Bezüglich der Anzahl der benötigten Regeln findet die sogenannte kombinatorische Explosion statt. Insbesondere bei steigender Anzahl der Regeln ist die Wissensbasis in Bezug auf die Abhängigkeiten zwischen den Regeln zunehmend weniger überschaubar. Auch die Erweiterbarkeit der Wissensbasis ist dann in Frage gestellt, da oft kaum bekannt ist wie sich die Aufnahme einer neuen Regel auswirkt. Die Praxis hat gezeigt, daß der erhoffte Erfolg im Einsatzgebiet der Fertigungssteuerung bislang ausgeblieben ist. Der wichtigste Grund liegt in der Notwendigkeit, den Entscheidungsprozeß eines Experten exakt situativ zu beschreiben. Angesichts der Tatsache, daß die Anzahl der Entscheidungssituationen aufgrund der Dynamik der Produktionsprozesse ins Unermeßliche steigt und der menschliche Disponent nur unter Unsicherheit seine Entscheidungen trifft, wird zur Lösung des Problems eine Methodik benötigt, die eine bessere Modellierung der Entscheidungsfindung menschlicher Disponenten erlaubt sowie die vorhandenen Unsicherheiten explizit repräsentiert und auswertet. Wie die Wissensbasis für den in diesem Aufsatz beschriebenen Ansatz ermittelt wird ist in 2.4 dargestellt.

1.5 Einsatz von Fuzzy–Logik

Dispositive Eingriffe in einem Teilbereich können, wie bereits geschildet, fast nie durch Algorithmen beschrieben werden, denn sie basieren auf teilweise langjährigen Erfahrungen. Es ist daher sinnvoll einen Fertigungsleitstand auf Meisterebene zu installieren, um den einem Fertigungsbereich zugeordneten Auftragspool optimal einlasten zu können. Die Aufgabe eines Fertigungsleitstandes sollte daher sein, den in der dezentralisierten Fertigungssteuerung tätigen Disponenten unter Berücksichtigung der in seinem Bereich vorhandenen Erfahrungen dergestalt zu unterstützen, daß ihm vom Computer Lösungsvorschläge angeboten werden. Dies erfordert eine Leitstandstruktur, in der der Bereichsdisponent beim Auftreten von Störgrößen sofort aktiv werden und ggf. eine neue Disposition ohne das PPS vornehmen kann. Nur so kann der Leitstand seiner Aufgabe als Regler gerecht und der Produktionsteilbereich insgesamt als Regelkreis betrachtet werden.

Es bietet sich an einen dezentral arbeitenden Fertigungsleitstand zu erstellen, der zum einen der Fertigungsstruktur angepaßt werden kann und zum anderen bei der Optimierung der Auftragsreihenfolge das situationsabhängige und spezifische Wissen des Disponenten berücksichtigt. Diese Forderungen rufen nach einem wissensbasierten System.

Die Klassifikation in diesem Artikel vorgestellten Systems ist in Abbildung 2 ersichtlich. Wobei nochmals herauszustellen ist, daß es sich um eine wissensbasiertes System und nicht um ein Expertensystem handelt. In dem Leitstandkern kommt eine zielorientierte Fertigungssteuerung auf Basis einer Fuzzy-Entscheidungsunterstützung zum Einsatz. Der Leitstand kann in bestehende Produktionsplanungssysteme integriert werden und dezentral die Auftragsreihenfolge automatisch optimieren. Der dezentral tätige Disponent hat weitreichende Eingriffs- und Entscheidungsmöglichkeiten, um das im Rechner vorhandene Modell der Fertigung mit seinem Wissen zu vervollständigen oder aktuelle Veränderungen zu berücksichtigen.

Kontrolliert wird die automatisch generierte Auftragsreihenfolge mit einem Simulator (siehe [4]), der Bestandteil des Leitstandes ist und auf die Daten im Rechner zurückgreift, so daß kein weiterer Bedienungsaufwand entsteht. Der Simulator entspricht dabei einer virtuellen Fertigung, mit dem der Produktionsablauf der Realität vorweggenommen werden kann. Ist der Disponent mit der generierten Auftragsreihenfolge aufgrund der Simulationsergebnisse (oder anderer Gründe) nicht einverstanden, hat er die Möglichkeit, die Entscheidungsgrundlage des Rechners zu analysieren und z.B. einzelne Zielkriterien zu überdenken und zu verändern. Danach kann dann wiederholt der Test mit dem Simulator erfolgen.

Diese Ansätze haben gemeinsam, daß sie auf durch den Menschen vorgebbaren Kennzahlen bzw. Zielwerten beruhen. Die verbale Beschreibung und das menschliche Verständnis dieser Werte kann von einem Computerprogramm nicht ohne weiteres nachvollzogen werden. Besonders qualitative Begriffe, wie *geringer Bestand, hohe Auslastung* usw., können von einem Computer mit den herkömmlichen Ansätzen nicht bearbeitet werden. Zu einem Problem wird dies vor allem dann, wenn Zielgrößen in wechselseitigen Beziehungen wie Konkurrenz, Behinderung oder Kooperation zueinander stehen. Die Ziele *geringer Bestand* und *geringe Durchlaufzeit* beispielsweise können unter verschiedenen Umständen miteinander konkurrieren oder kooperieren.

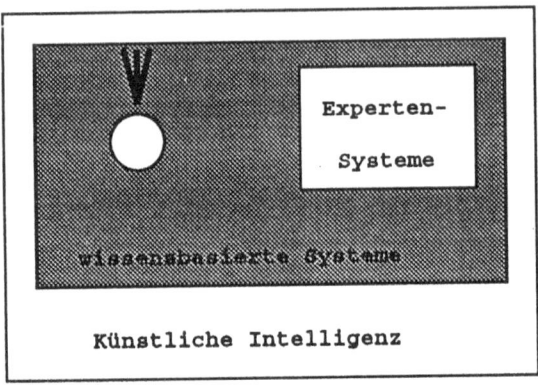

Abbildung 2: Systemeinordnung

Die Unschärfe der Information ist in der Regel durch mehrere Aspekte bedingt. In vielen Fällen ist das zu verarbeitende Wissen per Definition unscharf. Dies gilt z. B. dann, wenn das Wissen auf Abschätzungen beruht oder auf Messungen zurückzuführen ist. Schon die Tatsache, daß Meßfehler nicht auszuschließen sind, führt zu unsicheren Informationen. Ein weiterer wichtiger Grund für die Notwendigkeit, mit unscharfem Wissen umzugehen, ist die Tatsache, daß viele praxisrelevante Problemstellungen auf Grund ihrer Komplexität nur über den Weg einer geeigneten Abstraktion in den Griff zu bekommen sind. Abstraktion aber bedeutet einen bewußten Verzicht auf überflüssige Detailinformationen zugunsten einer vereinfachten Modellvorstellung über das betreffende Anwendungsgebiet.

Darüber hinaus sind häufig Probleme zu lösen, obwohl zum aktuellen Zeitpunkt ein Informationsdefizit vorliegt. In solchen Situationen gehen menschliche Entscheidungsträger aufbauend auf ihren Erfahrungen intuitiv vor und verwenden Daumenregeln und Vereinfachungen, die als solche unscharf sind. Die dabei entstehende Unschärfe wird nicht nur in Kauf genommen, sondern bewußt in den Prozeß der Entscheidungsfindung einbezogen.

1.6 Struktur der zielorientierten Auftragsreihenfolgeoptimierung

Die vorliegende Aufgabenstellung der Reihenfolgeoptimierung muß ein Mehrfachzielsetzungsproblem lösen, wobei zusätzlich die Schwierigkeit auftreten kann, daß einzelne Ziele unterschiedliche Prioritäten besitzen und sich teilweise auch widersprechen. Diese Beziehungen zwischen den Zielen sollten bei der Entscheidungsfindung berücksichtigt werden. Abbildung 3 stellt den Ablauf der zielorientierten Auftragsreihenfolgeoptimierung dar und Abbildung 4 die Struktur der Entscheidungstheorie (siehe auch [5]).

Der Entscheidungstheorie [2] liegt ein kognitives Modell des Entscheidungsprozesses zugrunde, das weitgehend an die Vorgehensweise menschlicher Entscheidungsträger angelehnt ist. Der Tatsache, daß menschliches Erfahrungswissen in komplexen Problemstellungen überwiegend auf Abschätzungen und Daumenregeln und somit auf unscharfen und unvollständigen Informationen beruht, wird in dieser Entscheidungstheorie durch die Modellierung der Zusammenhänge mit Hilfe von *Fuzzy-Mengen*, *Fuzzy-Relationen* und *Fuzzy-Prädikaten* Rechnung getragen. Die Verknüpfung einzelner Aussagen und Entscheidungen geschieht unter Rückgriff auf die Fuzzy-Logik.

Abbildung 3: Zielorientierte Auftragsreihenfolgeoptimierung

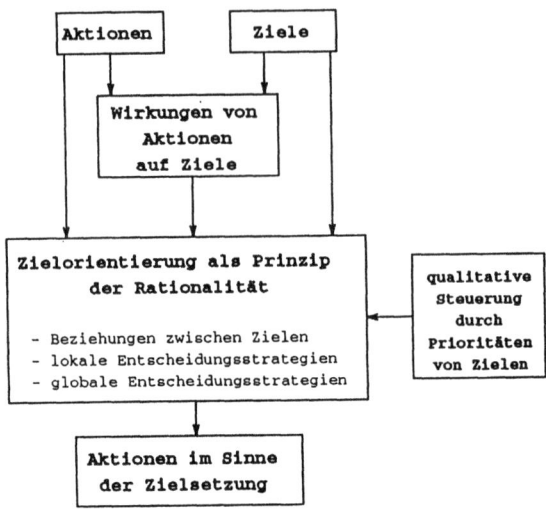

Abbildung 4: Struktur der Entscheidungstheorie

Die Vorgehensweise lehnt sich an das Verhalten des Diponenten an. Für den Fall, daß eine komplexe Entscheidung getroffen werden muß, verschafft sich der Disponent einen Überblick über die ihm zur Verfügung stehenden Maßnahmen (Aktionen) und die Ziele, die er erreichen möchte. Davon ausgehend schätzt er die Wirkungen der Aktionen auf die Ziele ab, wobei Wechselwirkungen und Beeinflussungen untereinander berücksichtigt werden müssen. Der Disponent wird immer die Aktionen auswählen, die seine Ziele am besten unterstützen. Hierbei sind Prioritäten der einzelnen Ziele zu berücksichtigen, die sich in unterschiedlichen Entscheidungssituationen ggf. verändern können. Die Wirkungen der Aktionen auf Ziele sind in sogenannten Wirkungsmengen beschrieben (siehe Abschnitt 2) und bilden den wesentlichen Teil der Wissensbasis. Sie liegen in der Regel als Schätzwerte vor. Dieser Sachverhalt wird durch die Verwendung von *Unscharfen-Mengen* berücksichtigt.

2 Wissensbasis

Die Wissensbasis enthält folgende Hauptbestandteile:

1. Ziele
2. Beziehungen zwischen Zielen
3. Aktionen
4. Wirkungen von Aktionen auf Ziele
5. Bewertungsfunktionen zur Wirkungsmengenbestimmung

Im folgenden werden die Bestandteile der Wissensbasis eingehender beschrieben.

2.1 Ziele

Die *Wissensbasis* enthält Informationen über die *Ziele* der Werkstatt. Für die Berechnung einer Auftragsreihenfolge müssen jedem Ziel *Prioritäten* aus einem Bereich von [0..1] zugeordnet werden. Diese Prioritäten werden den Zielen in der Wissensbasis zugeordnet. Je höher die Priorität eines Zieles, desto wichtiger die Einhaltung, während Ziele mit niedriger Priorität eher vernachlässigt werden können.

Diese Prioritäten bestimmen entscheidend die Auswahl der *Aktionen*, da die Auswahl einer Aktion immer *Wirkungen* auf das Erreichen von Zielen hat. Wichtig ist auch das Wissen um die *Beziehungen* zwischen Zielen. Ziele können sich in unterschiedlicher Art und Weise unterstützen, behindern oder auch nicht beeinflussen. Diese Beziehungen zwischen Zielen müssen systemabhängig abgeleitet und formuliert werden, z.B. : „Das Ziel Rüstzeitminimierung *behindert schwach* das Ziel Termintreue". Diese Beziehungen müssen für jedes Zielpaar aufgestellt werden, in beide Richtungen, da bei obigem Beispiel die priorisierte Verfolgung des Ziels Termintreue nicht unbedingt das Ziel Rüstzeitminimierung beeinflußt.

2.2 Aktionen

Grundlage der Auftragsreihenfolgeoptimierung ist ein Scheduling-Algorithmus (hier Force Directed Scheduling = FDS [9]), der während seines Ablaufes in die Situation kommt, Entscheidungen treffen zu müssen und somit über mehrere Freiheitsgrade verfügt. Der zugrundeliegende FDS-Scheduling-Algorithmus hat mehrere Freiheitsgrade.

1. Der *Einfluß* der Zuordnung eines Bearbeitungsvorganges auf eine Maschine M zum Zeitpunkt t wird mit Hilfe einer Formel berechnet. Diese Formel berechnet die Wahrscheinlichkeit, mit der diese Maschine zu diesem Zeitpunkt belegt ist. Je geringer die Wahrscheinlichkeit, desto geringer die Konkurrenz zwischen Bearbeitungsaufträgen an dieser Maschine zu diesem Zeitpunkt. An dieser Stelle kann die *Entscheidungstheorie* eingesetzt werden, um diese Formel in Abhängigkeit von der Zielsetzung zu modifizieren. Es stehen mehrere *Terme* zur Auswahl, mit der die Formel ergänzt werden kann. Diese Terme modifizieren die Formel zur Berechnung des Einflusses derart, daß die Ziele der Werkstatt unterschiedlich unterstützt werden.
Die Auswahl eines dieser Terme stellt an dieser Stelle eine *Aktion* dar. Folglich gibt es an dieser Stelle genau so viele Aktionen, wie Terme zur Modifikation der „Einfluß-Formel" zur Verfügung stehen. Jede Aktion, hier Auswahl eines Terms, hat Einfluß auf das Erreichen von Zielen, im weiteren *Wirkung* genannt.

2. Der zweite betrachtete Freiheitsgrad existiert an der Stelle, wo für mehr als einen Bearbeitungsvorgang für Maschine M und Zeitpunkt t der gleiche *Einfluß-Wert* berechnet wurde. An dieser Stelle muß entschieden werden, welcher Bearbeitungsvorgang zugeordnet werden soll.
Hier stellt die Zuordnung eines Bearbeitungsvorganges eine Aktion dar. Auch hier gilt wieder, daß die Auswahl einer Aktion Einfluß auf die Ziele hat. Mit Hilfe der *Entscheidungstheorie* wird hier die Aktion ausgewählt, die den günstigsten Einfluß auf die Verfolgung aller Ziele abhängig von deren Prioritäten hat.

Die Aktionen eines Freiheitsgrades werden zu einer *Klasse* (Menge) zusammengefaßt.

2.3 Wirkungen von Aktionen auf Ziele

In den vorherigen Abschnitten wurden die Bedeutung von Zielen und Aktionen erläutert. Es wurde bereits angesprochen, daß die Auswahl einer Aktion das Erreichen eines oder mehrerer Ziele beeinflußt. Im folgenden wird von der *Wirkung* von Aktionen auf Ziele gesprochen. Die Wirkung von Aktionen auf Ziele wird in der *Wirkungsmenge* zusammengefaßt.

Bei vier Zielen, z.B. Rüstzeit, Bestand, Auslastung und Termintreue, besteht die Wirkungsmenge einer Aktion aus vier Einträgen. Wirkungsmenge einer Aktion x : $W_x = \{R_x, B_x, A_x, Tx\}$ [1]. Die Elemente der Wirkungsmenge enthalten Werte zwischen [-1..1]. Positive Werte drücken Unterstützung, negative Werte Behinderung eines Zieles aus.

Die Wirkungsmengen einer Aktion müssen durch Testläufe und/oder *Bewertungsfunktionen* ermittelt werden. Insbesondere bei der Klasse von Aktionen zur Zuordnung eines Bearbeitungsvorganges müssen die Wirkungen bezüglich Rüstzeitminimierung und Bestandsminimierung dynamisch berechnet werden, da diese in Abhängigkeit vom Vorgänger auf dieser Maschine betrachtet werden müssen. Zur Berechnung dieser Wirkungen eignen sich besonders Bewertungsfunktionen, die z.B. die aktuell anfallende Rüstzeit in einen adäquaten Wirkungwert transformiert. In Abbildung 5 ist ein Beispiel für eine Bewertungsfunktion angegeben.

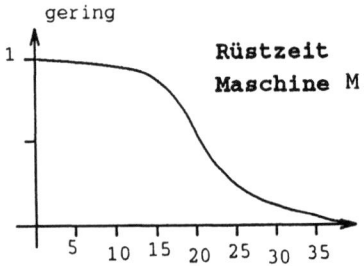

Abbildung 5: Rüstzeit Maschine M

2.4 Automatische Wissensaquisition

Um die oben beschriebenen Probleme der Akquisition eines Basiswissens zu umgehen sieht dieser Ansatz der Fertigungssteuerung mit Fuzzy-Logic eine automatische Konfiguration der Wissensbasis auf der Basis eines Simulationsmodells vor. Der Ablauf ist in

---
[1] R_x entspricht Wirkung auf Ziel Rüstzeitminimierung
B_x entspricht Wirkung auf Ziel Bestandsminimierung
A_x entspricht Wirkung auf Ziel Auslastung
T_x entspricht Wirkung auf Ziel Termintreue

Abbildung 6 beschrieben. Ausgehend von einem Auftragspool, den Arbeitsplänen, den Stücklisten, Transportzeiten zwischen einzelnen Bearbeitungsgängen, den Verfügbarkeiten der Maschinen, dem Wissen und Zielen des Disponenten wird automatisch ein detailliertes und ereignisorientiert ablaufendes Simulationsmodell der Fertigung generiert. Dieses Simulationsmodell liefert sämtliche Daten, die für die Beurteilung der Qualität einer Auftragsreihenfolge wichtig sind.

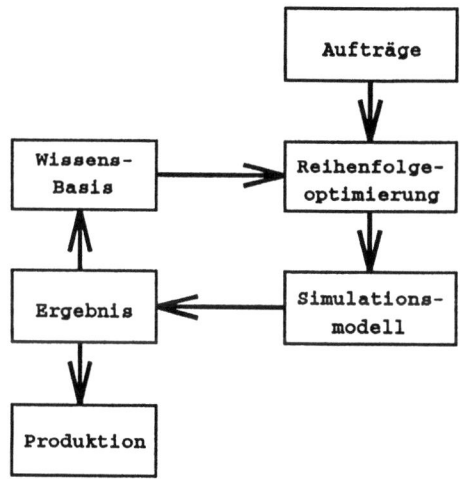

Abbildung 6: Wissensakquisition

Der Ablauf für die Ermittlung einer Wissenbasis sieht wie folgt aus (siehe Abbildung 6). Mittels der zielorientierten Auftragsreihenfolgeoptimierung wird auf Basis einer berechneten Auftragsreihenfolge ein Simulationsmodell erzeugt. Der Disponent definiert seine Ziele und die Beziehungen, die er erfahrungsgemäß zwischen diesen Zielen sieht, auf der Basis linguistischer Variablen. Mit diesen Daten wird ein Simulationslauf durchgeführt, der die Aufgabe hat die Wirkungen von Aktionen auf die Ziele zu berechnen. Befindet der Disponent das Simulationsergebnis für gut, dann bleibt die Wissensbasis in dieser Form erhalten ansonsten kann eine erneute Auftragsreihenfolgeplanung durchgeführt und erneut mit der Simulation überprüft werden. Hat der Disponent für eine bestimmte Werkstattkonfiguration eine seinen Anforderungen genügende Auftragsreihenfolge ermittelt, dann werden die zugrundeliegenden Wirkungsmengen in die Wissensbasis aufgenommen und die nächste Auftragsreihenfolgeoptimierung kann falls gewünscht ohne den Disponenten vorgenommen werden. Die Eingriffsmöglichkeiten des Disponenten sind in Abbildung 7 dargestellt.

Der Disponent kann dieses Modell im Dialog und graphisch interaktiv den aktuellen Betriebsbedingungen anpassen, soweit die Daten nicht von der BDE erfaßt worden sind. Für die Ermittlung einer optimierten Auftragsreihenfolge in einem Teilbereich der Produktion ist dann nicht mehr ein starres Modell für den gesamten Betrieb vorhanden, sondern ein dem PPS untergeordnetes Instrument für die Reihenfolgeplanung, das

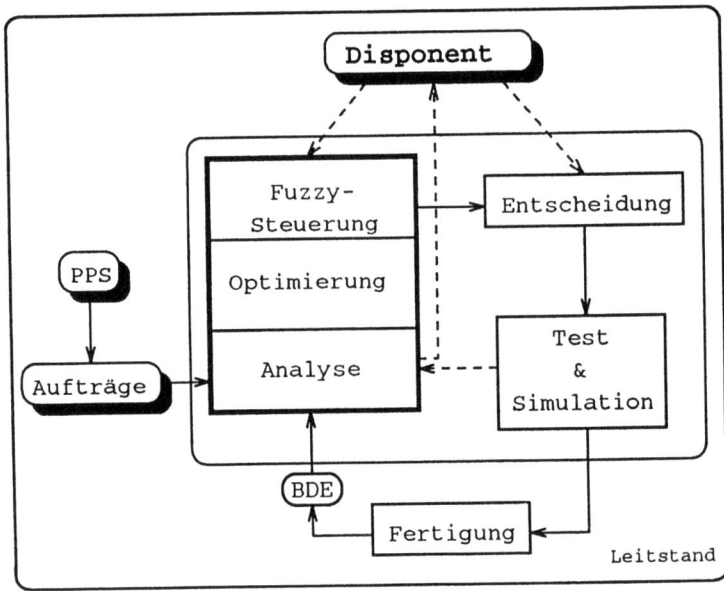

Abbildung 7: Eingriffsmöglichkeiten des Disponenten

spezielles heuristisches Wissen und individuelle Betriebskonstellationen berücksichtigen sowie kurzfristig Auftragsreihenfolgen optimieren kann.

3 Beispiel einer Zuordnung

An dieser Stelle soll die Vorgehensweise bei der Zuordnung von Bearbeitungsaufträgen an eine Maschine M zu einem Zeitpunkt t anhand eines Beispiels erläutert werden.

Die Ausgangssituation ist in Bild 8 zu erkennen. Die Zeilen der Matrix geben die Betrachtungszeitpunkte wieder, die Spalten markieren die Maschinen.

Es sei folgende Situation gegeben :

Auf Maschine $Ma1$ wird zum Zeitpunkt 1 der Bearbeitungsvorgang $B1$ bearbeitet. Während der Auftragsreihenfolgeplanung soll nun durch den FDS ein Bearbeitungsvorgang bestimmt werden, der danach zum Zeitpunkt 2 auf Maschine $Ma1$ laufen soll. Der FDS berechnet nun für alle in Frage kommenden Bearbeitungsaufträge[2] den Wert des Einflusses.

1. Es wird ein Bearbeitungsvorgang ermittelt, dessen *Einfluß-Wert* zu diesem Zeitpunkt der geringste ist. Dieser wird dann der betrachteten Maschine zugeordnet.

---

[2] nur die, die aufgrund ihres Zeitrahmens jetzt beabreitet werden könnten

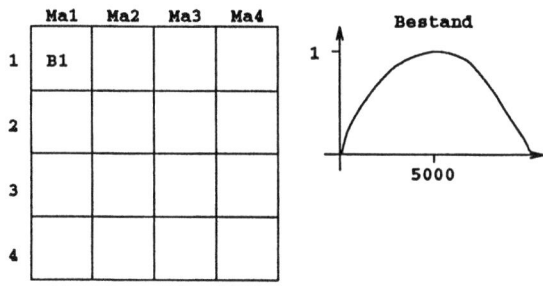

Abbildung 8: Beispiel einer Zuordnung

2. Zwei oder mehrere Bearbeitungsvorgänge haben zu diesem Zeitpunkt den gleichen geringsten *Einfluß-Wert*. Um zwischen diesen den Bearbeitungsvorgang auszuwählen, dessen Wirkungen den Zielprioritäten am besten entprechen, wird die Entscheidungstheorie (ET) benutzt.

Im Folgenden soll der zweite Fall näher beschrieben werden, da der Erste trivial ist. Zuerst einmal müssen die Ziele und ihre Prioritäten bekannt sein :

1. Ziel : Rüstzeitminimierung mit Priorität 0.8

2. Ziel : Bestandsminimierung mit Priorität 0.5

3. Ziel : gleichmäßige Auslastung mit Priorität 0.6

4. Ziel : Termintreue mit Priorität 0.8

Die Entscheidungstheorie braucht *Wissen* über die Wirkungen der Zuordnung eines Bearbeitungsvorganges. Dieses Wissen ist wie folgt repräsentiert :[3]

Name Bearbeitungsvorgang	Ziel 1	Ziel 2	Ziel 3	Ziel 4
B2	R_{B2}	B_{B2}	A_{B2}	T_{B2}
B3	R_{B3}	B_{B3}	A_{B3}	T_{B3}

Die Variablen (R_{B2}, B_{B3} usw.) in den Zeilen stehen für die Rückgabewerte von maschinenabhängigen Bewertungsfunktionen. Diese Bewertungsfunktionen wandeln die zu erwartende Rüstzeit etc. in Fuzzy-Werte um, die dann die Wirkung der Auswahl eines Bearbeitungsvorganges B_i auf die Ziele angeben. Mit diesen Eingaben (Zielprioritäten, Aktionen und Wirkungen) kann die Entscheidungstheorie aus mehreren Kandidaten den besten Auswählen.

Angenommen, nach Auswertung der Bewertungsfunktionen ergibt sich folgendes Bild:

---

[3] Annahme, es stehen zwei Bearbeitungsvorgänge B2,B3 in diesem Zeitfenster zur Auswahl

Name Bearbeitungsvorgang	Ziel 1	Ziel 2	Ziel 3	Ziel 4
B2	0.8	0.5	0.5	0.6
B3	0.7	0.5	0.6	0.7

Dies bedeutet, daß B2 das Ziel Rüstzeitminimierung besser Unterstützt als B3, B3 jedoch auch die Ziele Auslastung und Termintreue gut unterstützt. Die Beziehungen zwischen den Zielen unter Berücksichtigung ihrer Prioritäten müssen überprüft werden und gegebenenfalls entschieden werden, welche Ziele zu vernachlässigen sind. Danach muß zwischen B2 und B3 aufgrund ihrer Wirkungsmengen entschieden werden. Falls alle Ziele unabhängig voneinander sind, würde sicherlich B3 ausgewählt werden, da er insgesamt alle Ziele am besten unterstützt.

Literatur

[1] **H. Glaser** Verfahren zur Fertigsteuerung in alternativen PPS-Systemen - Eine kritische Analyse, in CIM-Management Heft 2/92

[2] **R. Felix** Entscheidungen bei qualitativen Zielen, Dissertation, Dortmund 1991

[3] **R. Felix** reasoning with Uncertainty in the Knwoledge Engineering Environment KEE, Developing and Managing Expert Systems Programs, Waschington DC, 1991

[4] **U.Schmidt** DOSIMIS-3 Benutzerhandbuch, Dortmund 1992

[5] **A. Adelhof, R.Felix, M. Kleiner** Modellierung von zielorientierten Entscheidungen am Beispiel eines Biegeverfahrens, VDI-Z Nr.9, 1991

[6] **SDZ-GmbH** SimAL–Simulationssystem für die Auftragsreihenfolgeplanung und Losgrößenbestimmung, Dortmund 1991

[7] **H.P. Wiendahl** Belastungsorientierte Fertigungssteuerung, Carl Hanser Verlag München Wien 1987

[8] **Ploentzke Informatik** Marktspiegel Fertigungsleitstände 1991

[9] **P. Paulin, J. Knight** Force Directed Scheduling, in Automatic Data Path Synthesis, 24th Design Automation Conference, 1987

Fuzzy Control Research at Siemens Corporate R&D*

Hans Hellendoorn   Michael Reinfrank

Abstract

Siemens Corporate R&D has decided to undertake a significant research effort in the field of fuzzy control. Both theoretical and practical aspects of fuzzy control and related areas in fuzzy set theory will be extensively researched in close connection with several universities, research institutes and industrial companies.

Motivation and Introduction

The two main reasons for Siemens's involvement into the area of fuzzy control are:

- The state-of-the-art in fuzzy control reveals clearly that there is a sufficiently wide range of applications where fuzzy control provides results superior to conventional control techniques. Fuzzy control systems are often characterized by their robustness, easy maintainability, and their ability to achieve good controls with comparatively low development and implementation efforts and costs.

- In a wider perspective fuzzy control appears to be very useful when applied to the identification and control of ill-structured systems, where e.g. linearity and time invariance cannot be assumed, the process is characterized by significant transport lags, and is subject to random disturbances. These kinds of systems are difficult to model from a conventional point of view.

However the lesson from the state-of-the-art in fuzzy control shows that most of the present applications of fuzzy control systems can be exemplified by relatively simple fuzzy control devices embedded in e.g. home appliances, air conditioners, cars, audio and video products, etc. We do not believe that fuzzy control when applied

---

*Published in: R. Kruse & P. Siegel (eds.), *Symbolic and Quantitative Approaches to Uncertainty*, (Proceedings of the European Conference ECSQAU, Marseille, France, October 1991) Berlin, Springer Verlag, pp. 206–210. Also appeared in: *Proceedings of the IJCAI-91 Workshop on Fuzzy Control*, Sydney, 1991, pp. 41–45. Will also appear in: *Proceedings of the Fuzzy Control Workshop*, Texas, Austin, 1991.

to these kinds of problems will completely replace conventional control techniques. Therefore, our strategy is to use it only when it actually has better performance, and provides for an easy integration with conventional approaches.

On the other hand, when looked upon from a wider perspective further research in fuzzy control can provide the formal tools for modelling and using the knowledge of an experienced process operator needed to cope with the control of ill-structured processes. For example, in a lot of cases there is a lack of a well-posed mathematical model. At the same time, the ability of the operator to deal with such a process is recognized and his knowledge of control actions can be described linguistically as a set of rules. Secondly, the human understanding of the process and its conventional mathematical description is alien, and this results in a lack of effective man-machine interface.

Applications of fuzzy logic usually fall into one of the three classes control systems, classification systems, and expert systems.

- Control such as industrial plant control or air conditioning control,

- Classification such as character recognition or situation assessment in environmental protection,

- Expert systems in the classical domains of diagnosis, design and planning.

While in the latter case *fuzzy* expert systems compete with conventional AI-based expert systems and promise only marginal improvements, we expect major advances in the areas of fuzzy control and classification.[1]

Issues in Fuzzy R&D

While for simple applications of fuzzy control the basic technology is available, and the main problems concern efficient design and implementation methods (see below), broadening the scope of the applications to the automation of very complex and ill-understood systems requires more fundamental research and may only be realized in a long term perspective. Some research topics in our current focus of interest are:

- A declarative, logical specification of fuzzy controllers [1] that so far are only described in a procedural way and in particular also formal description methods for the closed system including the fuzzy controller and the process. Our expectation is that, for example, formal stability criteria can be more easily achieved on the basis of such a formalism.

---

[1]In particular, compared to other uncertainty calculi, fuzzy logic does not offer novel or better solutions to the problems due to long inference chains that usually occur in classical expert systems. On the other hand, in fuzzy control and classification the inference chains are usually short and the main problem often is to evaluate a large number of parameters with informal decision criteria, which is a particular strength of fuzzy set theory.

- A systematic combination of conventional and fuzzy control techniques. A first step towards this direction is Palm's fuzzy sliding mode controller [2].
- Results of this work are expected to contribute also to better design methods for fuzzy controllers. Another main area of our interest is self-organizing controllers based on machine learning methods including neural-net based approaches [3].

Besides these research activities, the direct application of some readily available fuzzy control techniques and tools (software and hardware) is of course also of particular interest for a large number of operating groups in Siemens. However, there are a number of difficulties when trying to apply the existing methods and tools for concrete applications, viz.

- The present generation of control engineers are not familiar at all with the theories of fuzzy control, neither have they any practical experience in applying existing techniques and tools.
- There are no courses and text-books available at any level of education (undergraduate, postgraduate and on-site training).
- There is no systematic knowledge of what kind of fuzzy control corresponds to what kind of control problems.
- There is no general design methodology for the development and implementation of fuzzy controllers.

The last two points show that currently the development of fuzzy control systems is largely a trial and error process. What aggreviates the problem even further is the lack of good software support. These two major problems will be the focus of Siemens R&D in the first phase of the project. Our goal is to come up not with the universal tool for fuzzy control development, but with a set of efficient methods and tools that support the whole life-time of a fuzzy control application including

- Decision criteria: which type of fuzzy control is useful for which type of applications and what are the expected benefits. So far only local experience exists from isolated single applications.
- Design and specification methods.
- Simulation and test environments that are embedded into development packages for conventional control engineering.
- Evaluation criteria for the testing of the performance of fuzzy controllers. Since the philosophy of fuzzy control is often to achieve simple but robust systems with a low effort, evaluation criteria from classical control engineering are often inadequate.

Corporate Strategy and Cooperations

Siemens Corporate R&D has decided to act as both a user and a supplier of the fuzzy technology, and aspires to partial technological leadership in Europe. We think this is a realistic goal, because Siemens has the technological resources needed available in house, plus a significant research potential in the fields of AI, Control theory, and Computer science. So, for example, in the neighborhoud groups there are big projects in the fields of neural nets, qualitative physics, nonmonotonic reasoning and truth-maintainance systems. All these techniques lend themselves to be combined with fuzzy control, e.g. neural nets for self-learning fuzzy controllers, qualitative physics for temporal fuzzy control and truth-maintainance systems for the description of default values and non-monotonicity occuring in fuzzy controllers. Also, there is a wide variety of application areas covered by the Siemens operating groups to test the viability of the fuzzy control related techniques, methodologies, software and hardware products. Examples of these application are projects on home appliances, car electronics, hardware technology, *etc.*, but far more important for Siemens are future works on ill-structured problems, such as in environmental protection areas. At present, the number of researchers directly involved in the fuzzy control project at Siemens R&D itself is about fourteen, acting as a "Corporate Center of Competence," with sufficient funding for the next five years. In addition to that, there are several smaller, purely application-oriented activities spread over the Siemens operating groups.

In order to catalyse the development efforts, Siemens cooperates closely with Togai InfraLogic, Irvine, a Software House specialized in the field of fuzzy control.

. The research activities are embedded into several national and European joint efforts including a German government-funded project on fuzzy logic and its applications to control, classification, and expert systems. The rationale behind these projects is to gather a critical mass of researchers and to bring the substantial research potential available in Europe to bear on practical applications. The European activities are of course complemented by close contacts to American and Japanese institues such as the Berkeley Institute in Soft Computing BISC, and the Yokohama-based LIFE.

Summary

In summary, our goals are:

- To bring the state-of-the-art in fuzzy control into Siemens. Therefore, existing software and hardware tools are tested on a number of pilot applications, in cooperation with operating groups from Siemens, together with external partners.

- Developing a full range of software/hardware products for the development of fuzzy controllers of different types and purposes.

- To study some fundamental issues in fuzzy control so that hard problems in modern control theory can be successfully tackled with fuzzy means.

Finally, we think it is important to provide serious and realistic information to industry and to the public about the principles and the application potential of fuzzy logic, in contrast to the now predominating type of newspaper articles, which use oversimplified presentations and trigger by far exaggerated expectations. Besides intensive discussions with companion disciplines, such as control engineering or artificial intelligence, and, of course, training for engineers and scientists, this kind of "PR-measures" will be crucial if fuzzy control shall be a technique to stay. Its promises for advanced problem solving make it worthwhile trying.

References

[1] Driankov, D. & Hellendoorn, H., "Towards a *Logic* for a Fuzzy Logic Controller," *Proceedings of the ESCQAU-Conference Marseille,* Berlin, Springer Verlag, 1991.

[2] Palm, R., "Fuzzy Controller for a Sensor Guided Robot Manipulator," *Fuzzy Sets and Systems,* 31(1989)133–149.

[3] Hecht, A., Hellendoorn, H. & Leufke, A., *Machine Learning for Fuzzy Control,* forthcoming technical report of the "Siemens Intelligent Control Systems Laboratory," Munich, Germany, 1991.

Teil 3

Online Development Tools for Fuzzy Knowledge-Based Systems of Higher Order

C. von Altrock and B. Krause

INFORM GmbH
Pascalstraße 23
5100 Aachen, Federal Republic Germany

Keywords: Fuzzy Logic Control, Application, Advanced Fuzzy Technologies, Tools, On-Line-Development, Water Purification

Abstract:

Most fuzzy logic control applications today use fuzzy production rules to represent relations between input and output state space. With that, the controller's reaction can be formulated for a specific situation such as: IF angle=positive_large AND derivative_of_angle=negative_ small THEN set_value =large_ negative. Since the context in these rules is derived heuristically by the engineer, these systems are often presented as "fuzzy expert systems". While this technology has successfully been applied in a variety of control problems, the approach is limited to rather simple structures - rule sets in these systems almost never contain more than 50 rules. For the control of more complex processes which involve a great amount of human expertise, advanced technologies have been developed, such as Fuzzy Associative Maps (FAMs), compensatory fuzzy operators and the so-called "on-line" technology for the on-the-fly optimization of a fuzzy controller in real-time. These advanced fuzzy technologies are presented and their usage is demonstrated by a controller implemented in a waste water purification plant in Vienna, Austria.

1. Introduction

State-of-the-art in fuzzy logic control applications is the simple calculus of "fuzzy-if-then-rules" which predominantly use MIN/MAX operators. For the implementation, mainly precompiler-based tools are used. Though this concept is sufficient for simple control applications or consumer goods, it is often insufficient for complex control applications. This work investigates which advanced methods render the application of fuzzy control in complex problems possible and how fuzzy controllers may be built more efficiently. The results of this theoretical work has already been implemented in a professional fuzzy logic development system, the *fuzzy*TECH tool family [11].

2. Advanced Fuzzy Technologies for Complex Control

By standards of conventional control theory, most commercial applications of fuzzy control today are of a moderate complexity level. They use the simple technology of "Fuzzy-IF-THEN-rules", employing linguistic variables and defuzzification heuristics [4, 16, 20] even though more sophisticated reasoning and defuzzification methods are available in theory [8, 14, 13, 9]. Some of these advanced technologies have already been applied with very promising results [1]. If the control problem becomes more complex, this technology is insufficient. We now focus on three aspects: size of the rule base, fuzzy operators and inference strategy.

2.1 Size of the Rule Base

It is often claimed that rule-based systems are more lucid and easy to comprehend than procedural programs. This may hold true for small systems, but when designing fuzzy controllers for problems more complex than inverted pendulums, this may prove untrue. Consider a medium size rule base with 30 rules like:

```
IF ( ( Temp IS very_high ) AND NOT ( ( Press IS above_norm )
OR ( Antech_Press IS low ) ) AND ( O2_Frac IS NOT normal ) )
THEN ( CH4_Val IS throttled AND Carb IS low )
```

Therefore, the concept of a normalized rule block representation was introduced [2, 3]. For this concept, a graphic representation which allows even large rule bases to remain easy to comprehend exists. To allow for greater flexibility, the inference strategy is separated in "near inference" and "far inference" [2, 3]. The near inference strategy controls the evaluation of the rules within a rule block and the far inference strategy guides the information flow between them. Figure 2 shows the graphical representation of a rule block as a screenshot of *fuzzy*TECH [3]. If there are more than 2 input variables, those not displayed in the matrix may be selected in the list boxes in the lower part of the screen. For the later presented case of the purification plant controller, more than 240 rules were implemented.

2.2 Fuzzy Operators

In most known fuzzy control applications, MIN/MAX-operators are used to model the "and" and "or", respectively. Empirical research [15, 19] has shown that these operators are only a rough approximation of the linguistic meaning they have to represent. This also holds for fuzzy control applications.

Consider, for example, the following rule which has been taken from the controller of a chemical plant:

```
IF    Temperature_of_Combustor = high
AND   Pressure_of_Antechamber = above_med
THEN  Methane_Valve = throttled
```

For the plant, this rule throttles the methane flow if the state in the combustor becomes critical. The degree to which the current state of the plant is rated critical is computed by combining the degree to which the Temperature_of_Combustor is high and the degree to which the Pressure_of_Antechamber is above_medium with the operator chosen for the linguistic "AND". Consider the following possible states of the plant:

#	Temp.	μ_{high}	Press.	μ_{above_med}
1	920°C	0.3	40 bar	0.4
2	920°C	0.3	51 bar	0.8
3	910°C	0.2	69 bar	1.0

When using the minimum operator, state #1 and state #2 are rated equally critical by the fuzzy controller though state #2 would be rated more critical by an engineer. Comparing state #3 to state #1, the fuzzy system would rate state #1 as more critical while an engineer would rate state #3 as more critical. This is due to the fact that the minimum operator does not allow for the "compensation" present in human judgement. When humans combine two concepts with the linguistic symbol "and", usually "a much more of the one makes good for the other". For the same reason, the maximum operator is inappropriate as well.

There are two ways to solve this problem. One is to define more terms for the linguistic variables "Temperature_of_Combustor" and "Pressure_of_Antechamber" like "somewhat_high", "very_high" or, respectively, "slightly_above_med" and "right_above_med". With these new terms, a set of rules which better fits the engineer's concept of criticality of the combuster can be defined. This approach, however, has two serious disadvantages. First, increasing the number of terms enlarges the number of rules to be defined. This not only impedes the design process and obfuscates system structure but increases the computational effort required of the controller. Secondly, and more importantly, when more terms for a linguistic variable than considered in the developing engineer's concept of that variable must be defined, the system ceases to be easily interpretable.

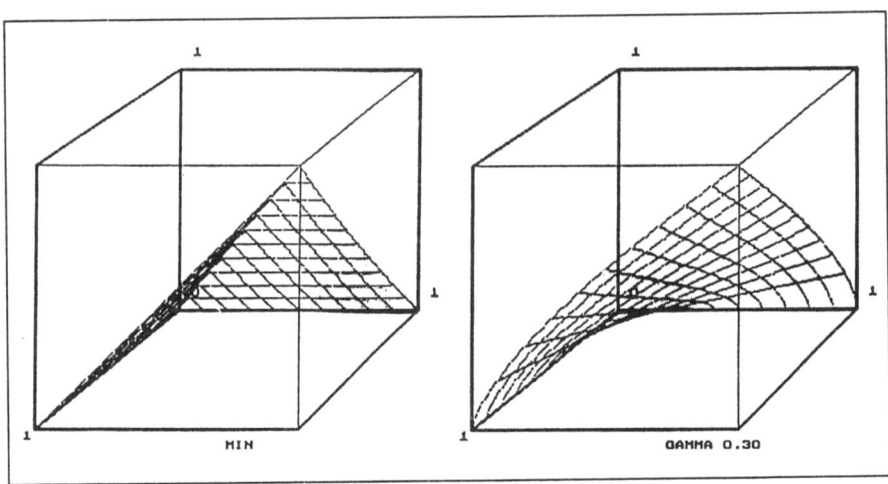

Fig. 1: Comparison of different fuzzy operators:
Left: Transfer Function of the Minimum-Operator.
Right: Transfer Function of the Gamma-Operator.
The horizontal axis shows the degree of truth of the two premises while the vertical axis shows the truth degree of the condition "both premises are true".

For the control of the purification plant, quite a few estimates similar to the presented example have to be made. For instance, nitrification processes in the clearing sump of the plant can be judged by measuring NO_3-nitrogen and NH_4-nitrogen ingredients. Only rating and comparing both parameters at the same time can establish whether or not nitrification is in a confident state. For this reason, so-called compensatory operators [20] have been implemented. With these type of operators, the degree of compensation may be tuned by a parameter. Figure 1 shows the aggregational properties of the minimum operator versus a compensatory operator with a γ (degree of compensation) of 0.3.

2.3 Inference Strategy

When the first prototype of a fuzzy controller is set up, one defines an initial set of rules which come to an engineer's mind as being "plausible" by his apriori concept of how to control the plant. Refining the control strategy in the optimization step often requires the deletion of certain rules and the addition of others. After a while, the stage is reached where formulating a rule or not is too coarse a means to further optimize a control strategy. Basically, in this stage of the optimization process, one winds up with a board of switches each representing a single rule. Switching on means defining a rule for a specific combination of input and output conditions; switching off means deleting this rule. This not only negates the philosophy of fuzzy logic -- since a switch is dichotomous -- but also makes optimization pretty much a trial-and-error task.

For a finer tuning means, some engineers have tried to tune the membership functions so that the desired behaviour is established. This clearly violates the concept of fuzzy control. The membership functions must represent the engineer's linguistic concept of physical figures in a self-explanatory nature. Fumbling around with the membership functions for the fine-tuning of a control strategy is therefore counter-productive, especially the more complex a problem becomes.

The solution is simple: if one wants to fine-tune the control strategy, one should not touch the membership functions, but tune the rules. This can be done by considering rules themselves as "fuzzy" by nature, not dichotomous.

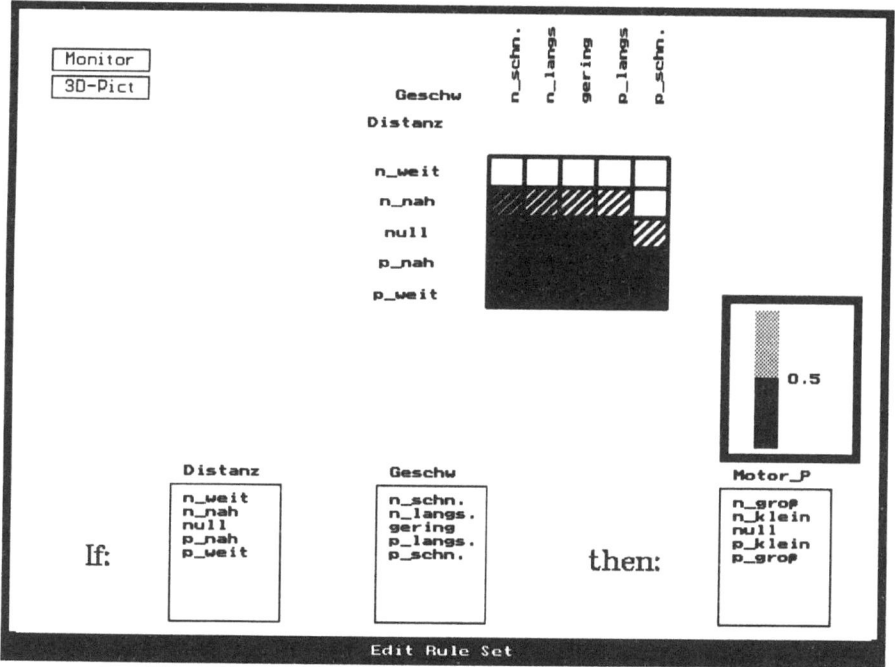

Fig. 2: Definition of Normalized Rule Sets as Matrixes (Screenshot of the fuzzyTECH-Shell).

One concept of such an advanced inference procedure was proposed by Zadeh [18] and is known as the "compositional rule of inference". This concept allows a "degree of support" (also called "degree of plausibility") to be associated with any rule. This concept also enables a very fine tuning of the rules but involves a large computational effort that forbids its usage in most real-time systems.

Another concept, so-called Fuzzy Associative Maps (FAMs) introduced by [kosko] and others, also allows for rules to be considered as "fuzzy". FAMs only require a small computational effort but only introduce a "weight" factor to each rule.

For both computational efficiency and appropriateness, *fuzzy*TECH supports a combination of these methods: the degree to which every rule fires is determined by applying a fuzzy operator to aggregate the degree to which the premise is fulfilled with its degree of support [2,3]. This can be interpreted as a "weight" for every rule. This method is rather simple to use: first, define degrees of support of only either zero or one. Second, for fine tuning, use values between 0 and 1. Figure 2 shows an example of such a definition of a rule matrix. The degree of support is indicated by the shade of gray [11].

3. A Development Approach for Complex Control Systems

Although fuzzy control systems are sometimes programmed in a conventional programming language, it is reasonable to use a tool which prevents repeated programming fuzzy methods like inference strategies or defuzzification methods. Special hardware like fuzzy processors can be used if a high-speed controller has to be set up.

For the definition of a fuzzy control strategy, a variety of software tools based on the concept of compiling/precompiling which already exist in the market can be used. Although this concept works well for the development of conventional software, it has several inherent drawbacks for the construction of fuzzy control systems.

At first one has to define an initial control strategy. This control strategy has to be compiled and linked to the process or its simulation for testing. If at first the controller does not work perfectly -- which is often the case -- the control strategy must be revised. For the revision, existing debuggers are clearly inappropriate since they either work on the compiled code or do not work in real-time. Additionally, when a fuzzy control strategy is being designed for a continuous process or its simulation, optimization is done by trying small definition changes and then analyzing the subsequent reaction of the control loop. To recompile the controller, and thereby interrupting the continuity of the process, any time a small change is made, development time is increased considerably.

For these reasons, an On-Line-Module has been integrated in the *fuzzy*TECH tool family. With this tool, a fuzzy logic control strategy can be graphically visualized while the fuzzy system is actually controlling the process in real-time. This enables the engineer to understand the dynamic behaviour of both the controller and the process.

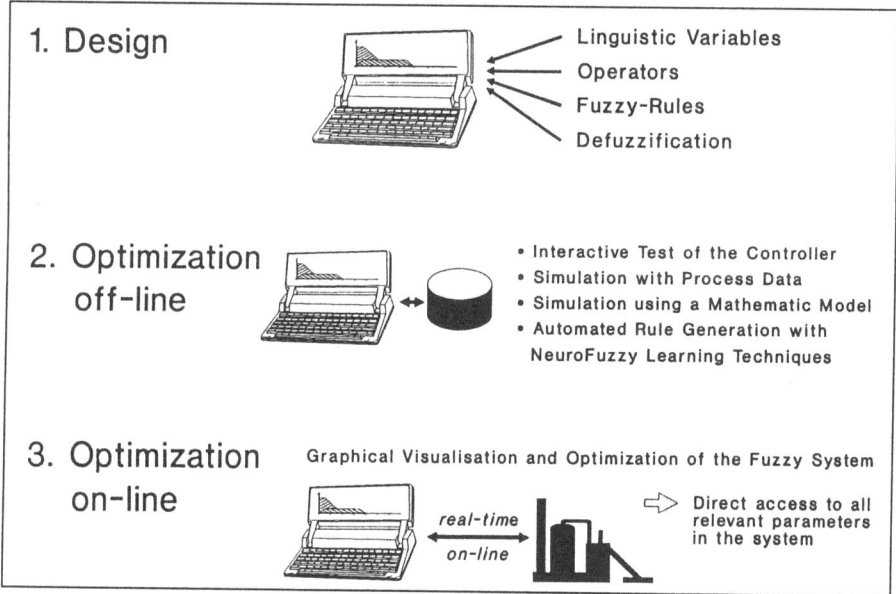

Fig. 3: *Development Approach for Complex Fuzzy Logic Control Systems [5]*
1. *Design: definition of linguistic variables, fuzzy operators, the fuzzy rule base and the defuzzification method. This step ends with a first prototype of the controller.*
2. *Off-Line Optimization: To check the controller's static performance, one can either test the controller interactively by applying input values and analyzing the information flow in the system or one can simulate the controller's performance on pre-recorded process data or a mathematical model of the plant, if available. This step ends with a refined prototype.*
3. *On-Line Optimization: The refined prototype is now optimized on the running process. This step establishes the final optimized system ready for implementation. To allow for on-line development, the workstation/PC running the fuzzyTECH-Shell is connected to the process controller hardware by just a serial cable.*

In the optimization step, one often wants to try out little modifications to the rule strategy or the membership functions to subsequently increase system performance. If a code-generating approach such as a fuzzy-precompiler is used, every time such a change is made, the controller has to be put off-line and the controller has to be recompiled. In addition to being very inefficient, this approach has another drawback: a continuous process is always being put out of its operating point by the recompilation. Hence, the effect of the control strategy modification can no longer be visualized by the engineer. This makes efficient optimization next to impossible. The water purification plant is a perfect example for such an application.

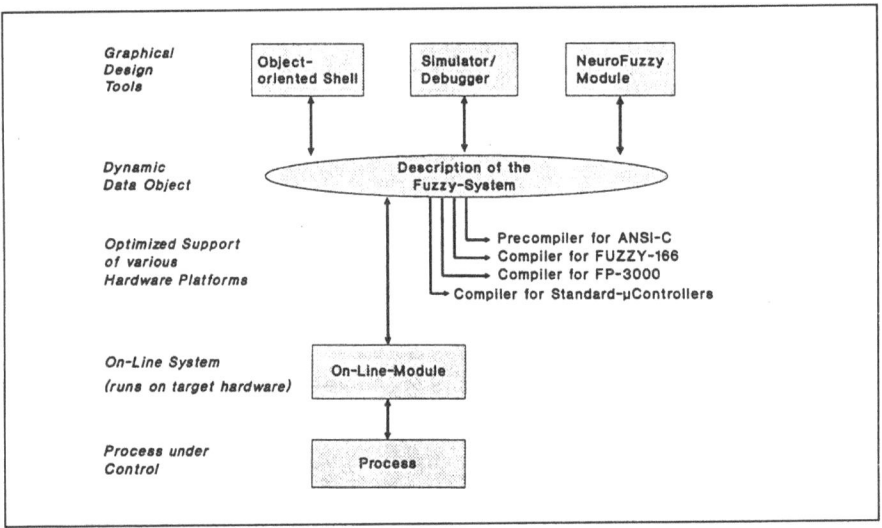

Fig. 4: Individual Tools of the fuzzyTECH family

Once the system is readily optimized, a precompiler or compiler can be used for a code-optimized implementation of the final system on the controller hardware. Figure 4 shows the different *fuzzy*TECH tools used for water purification controller development.

4. Experimental Results

Until now, the control of water purification plants has been done by human operators. The interdependence of biological, chemical and technical processes as well as the uncertainty about pollunt contents causes the ineffectual functionality of conventional control systems in the above-mentioned application.

The treatment of abnormal processing states is an especially important pitfall in conventional control systems and one that is often mishandled by human operators. Figure 5 shows the principle structure of a biological water purification plant.

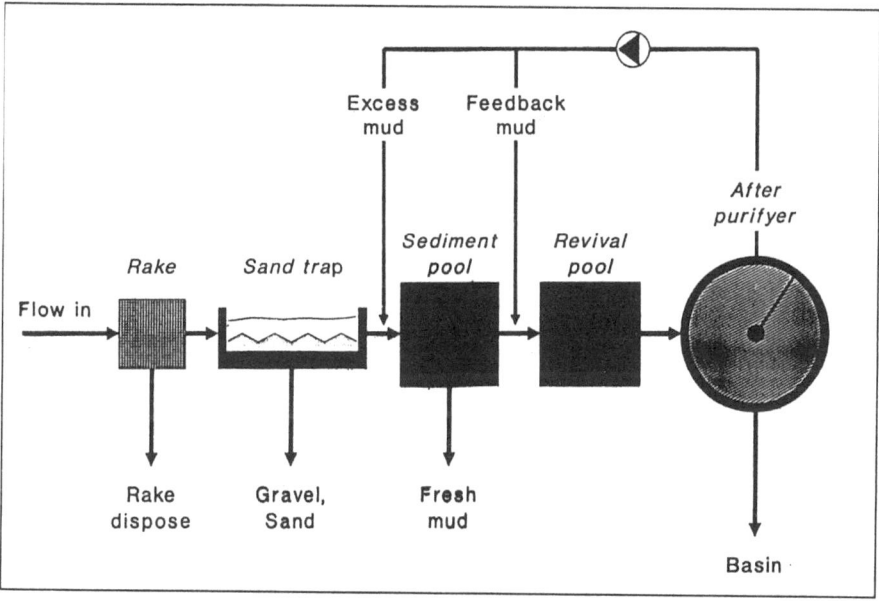

Fig. 5: Principle Structure of a Water Purification Plant

To support human operators in sludge plant control, information systems based on conventional information processing are used [5,7]. The lack of precise process information combined with the complex systems structure is the reason for the use of knowledge-based systems employing fuzzy set theory for knowledge representation [6,12].

In the given application, fuzzy components were integrated in a conventional information system [17]. The created fuzzy controller was built up to characterize abnormal operating states like the occurrence of bulking sludge in water purification plants. To avoid or repair these failures, different operation strategies must be taken into consideration. Choosing the best strategy depends at least on the correct classification of the failure situation in the current process state. This means that in addition to the actual process state, apriori known facts about plant configurations and industrial polluters have to be considered.

In the initial design step, system structure - imported and exported information including fuzzy- and defuzzification, linguistic variables, rule blocks, and fuzzy operators - was built up based on a three day knowledge acquisition with experts of process engineering.

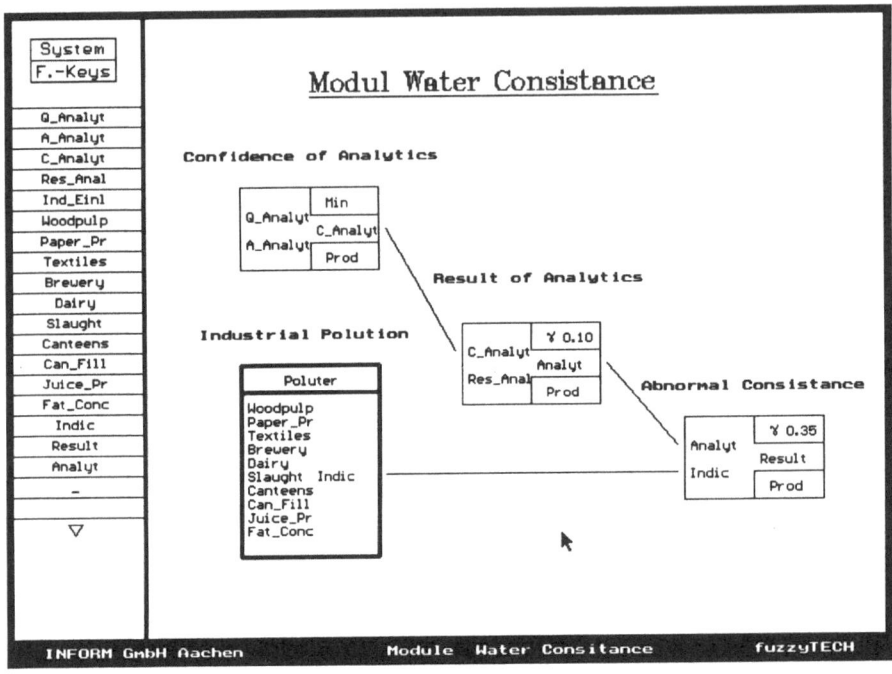

Fig. 6: Structure of a Fuzzy Component Indicating Abnormal Wastewater Consistency

Figure 6 shows the structure of one of the fuzzy components. This component handles on one hand with actual measured information - quality and result of a chemical analysis - and on the other with polluters information. As a result, the component indicates on abnormal wastewater consistency in case of bulking sludge.

The left side of figure 6 shows the list of linguistic variables. One of these variables representing the procentual amount of wastewater of one industrial polluter is shown in figure 7.

On the other side of the worksheet, the structure of far inference of the fuzzy system is depicted. In the first rule block, the confidence of the last available chemical analysis is calculated by its age and quality. In a second rule block, the results of the analysis are aggregated with the calculated measurement of confidence.

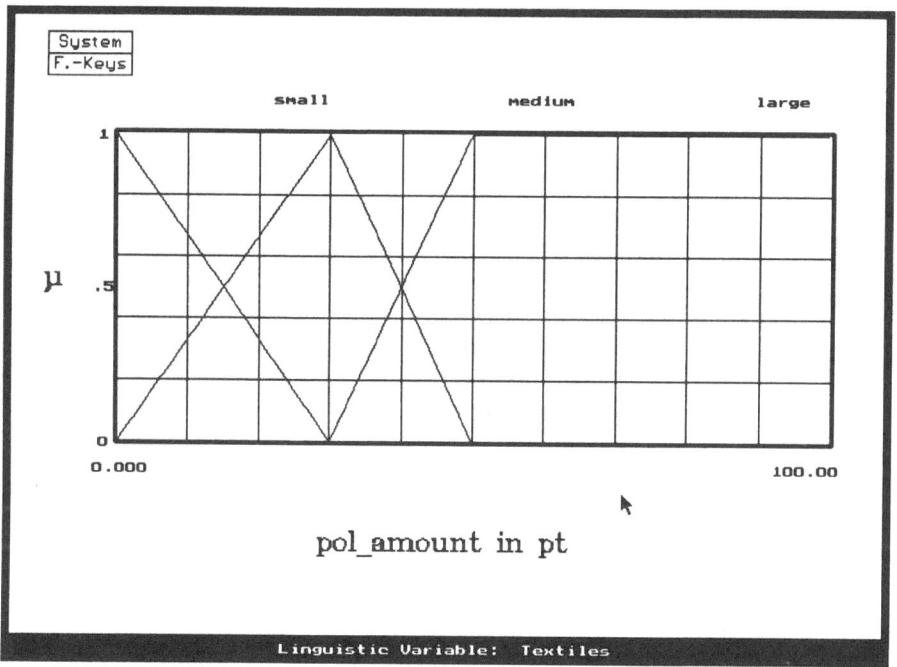

Fig. 7: Sreenshot of a Linguistic Variable Representing the Procentual Amount of Wastewater of One Industrial Polluter

The module named poluter considers the known industrial polution. As a consequence of both, the knowledge of polution of typical industries and results of chemical analysis, the abnormal water consistency in the clearing sump can be indicated or predicted and failure repairing strategies being started.

The prototype controller was refined in a second and third design step by off-line and on-line optimization.

References

[1] C. von Altrock, B. Krause and H.-J. Zimmermann, "Advanced Fuzzy Logic Control Technologies in Automotive Applications," Proceedings of 1992 IEEE International Conference on Fuzzy Systems, pp.835-843, San Diego, March 1992.

[2] C. von Altrock, B. Krause and H.-J. Zimmermann, "Framework of a Fuzzy Intelligence Research Shell," Working Paper 5/90, RWTH University of Aachen, Germany 1990.

[3] C. von Altrock, B. Krause and H.-J. Zimmermann, "Implementation of a Fuzzy Intelligence Research Shell," Working Paper 6/90, RWTH University of Aachen, Germany 1990.

[4] S. Assilian and E.H. Mamdani, "An experiment in Linguistic Synthesis with a Fuzzy Logic Controller," Internat. J. Man-Machine Stud. 7, pp. 1-13, 1975.

[5] S. Aya, J. Hirotsuji, S. Inoue, K. Maeda and M. Nonoyama, "A New Expert System Based on Deep Knowledge for Water and Wastewater Treatment Plant," in R. Briggs (Publ.), Instrumentation and Control of Water and Wastewater Treatment and Transport Systems, Proceedings of the 5th IAWPRC Workshop, Yokohama, pp. 219-226, 1990.

[6] M. B. Beck, A. Latten and R. M. Tong, "Fuzzy Control of the activated Sludge Process," Automatica, Vol 16, pp. 659-701, 1980.

[7] P. M. Berthouex and W. Lai, "Testing Expert System for the activated Sludge Process Control," Journal of Environmental Engineering, Vol. 116, pp. 890-909, 1990.

[8] J. C. Fodor, "On Fuzzy Implication Operators," FSS 42, pp. 293-300, 1991.

[9] R. Gorind and R. Zhao, "Defuzzification of Fuzzy Intervalls," FSS 43, pp. 45-55, 1991.

[10] M. M. Gupta and J. Qi, "Design of Fuzzy Logic Controllers Based on Generalized T-Operators," FSS 40, pp. 473-489, 1991.

[11] INFORM Software Corp., *fuzzy*TECH Reference Manual.

[12] D. M. Jonston, "Diagnosis of Watewater Treatment Processes," in Proceedings ASCE Speciality Conference, Computer Applications in Water Rescources, Buffalo, pp. 601-610, 1985.

[13] J. M. Keller and A. Nafarieh, "A new Approach to Inference in Approximate Reasoning," FSS 41, pp. 17-37, 1991.

[14] Mitsumoto and H.-J. Zimmermann, "Comparison of Fuzzy Reasoning Methods," FSS 8, pp. 253-285, 1992.

[15] U. Thole, H.-J. Zimmermann and P. Zysno, "On the Suitability of Minimum and Product Operators for the Intersection of Fuzzy Sets," FSS 2, pp. 173-186, 1975.

[16] R. M. Tong, "Analysis of Fuzzy Control Algorithms using the Relation Matrix," Internat. J. Man-Machine Stud. 8, pp. 679-686, 1976.

[17] E. Trenker, International Seminar on the Subject: Entwicklung und praktischer Einsatz eines komplexen Beratungssystems für Kläranlagen, Technical University Vienna, 1992.

[18] L. A. Zadeh, "Outline of a New Approach to the Analysis of Complex Systems and Decision Processes," IEEE Transactions on Systems, Man, and Cybernetics, Vol. SMC-3, No. 1, pp. 28-44, 1973.

[19] H.-J. Zimmermann and P. Zysno, "Latent Connectives in Human Decision Making," FSS 4, pp. 37-51, 1980.

[20] H.-J. Zimmermann, Fuzzy Set Theory - and its Applications, 2nd rev. Ed. Kluver, Boston, 1991.

Entwurf, Simulation und Implementierung von Fuzzy-Logic in einer integrierten Entwicklungsumgebung

Michael Hoffmann, TEDAS GmbH[1]

Zusammenfassung

Nach einem Überblick über die MATRIXx-Produktfamilie, die eine integrierte Umgebung für Entwurf, Simulation und Implementierung von komplexen Regelsystemen bietet, werden die Eigenschaften des Moduls RT/Fuzzy näher beschrieben. Es wird dargelegt, welche Methoden zur Behandlung unscharfer Logik angeboten werden und welche Optionen für Codegenerierung vorhanden sind.
Anhand eines im offenen Regelkreis instabilen Systems wird gezeigt, wie RT/Fuzzy gewinnbringend mit konventioneller Regelungstechnik kombiniert werden kann.

Design, Simulation and Implementation of Fuzzy Logic in one Integrated Environment

Problems which exceed an accurate mathematical description could have hardly or not at all been solved by classical control techniques. Now Fuzzy Logic promises an easier solution. Due to the increasing interest in this new control approach, appropriate design and implementation tools are more and more required. Furthermore, engineers wish to validate the performance of Fuzzy Logic by simulations during the design phase already.
This paper introduces RT/Fuzzy, a module of the MATRIXx product family. RT/Fuzzy is fully integrated in the MATRIXx environment so that the design process of complex control systems is continuously supported from the very beginning. In contrast to other fuzzy design tools, Fuzzy Logic can be combined with traditional control elements via a block diagram editor. Thus simulations can be used in the design phase already in order to evaluate the performance. A code generator serves to transform the block diagram description into real time C, Fortran or Ada programs.
Discussing an application example, it is demonstrated how Fuzzy Logic can be successfully combined with traditional control techniques.

1. Einleitung

Während Probleme, die sich einer genauen mathematischen Beschreibung entziehen, bisher nur schwer oder überhaupt nicht mit der klassischen Regelungstechnik bewältigt werden konnten, verspricht nun Fuzzy Logic eine einfachere Lösungsmöglichkeit. Mit dem zunehmenden Interesse an diesem neuen Regelungsansatz wächst auch die Nachfrage nach geeigneten Entwicklungs- und Implementierungswerkzeugen und

---

[1] Anschrift: Universitätsstraße 51, W-3550 Marburg/Lahn, Tel.: 06421/26077

der Wunsch, bereits in der Entwurfsphase durch Simulation die Leistungsfähigkeit der "unscharfen" Logik beurteilen zu können.

Im vorliegenden Beitrag wird das Modul RT/Fuzzy der MATRIXx-Produktfamilie vorgestellt. Eingebettet in die MATRIXx-Umgebung wird eine durchgängige Unterstützung während des Entwicklungsprozesses von komplexen Regelsystemen angeboten. Im Gegensatz zu anderen Fuzzy-Entwicklungswerkzeugen kann hier Fuzzy Logic graphisch interaktiv über einen Blockdiagrammeditor mit Elementen klassischer Regelungstechnik kombiniert werden. Dies ermöglicht auch, daß bereits im Entwurfsstadium über Simulationsrechnungen Aussagen zur Leistungsfähigkeit getroffen werden können. Ein Codegenerator gestattet die Umsetzung von der Blockdiagrammbeschreibung in echtzeitfähige C-, Fortran- oder Ada-Programme.

Im folgenden werden nach einem Überblick über die MATRIXx-Produktfamilie die Eigenschaften des RT/Fuzzy-Moduls näher erläutert. Als Anwendungsbeispiel wird ein - im offenen Regelkreis instabiles - System besprochen, für das zwei unterschiedliche Regelungsansätze realisiert wurden: a) Regelung nach konventionellem Ansatz und b) eine Kombination von klassischen Elementen und Fuzzy Logic.

2. Kurzbeschreibung der MATRIXx-Produktfamilie

Die MATRIXx-Produktfamilie von Integrated Systems Inc. bietet eine durchgängige Unterstützung während des Entwicklungsprozesses von komplexen Regelsystemen, angefangen bei der Systemmodellierung bis hin zum Prototypenversuch, Bild 1. Die verschieden Module lassen sich in vier Gruppen einteilen, Bild 2.

2.1 Systemanalyse und Reglerentwurf

Der MATRIXx-Kern erlaubt die Durchführung verschiedener Matrizenoperationen, das Erstellen von Grafiken und die Implementierung von benutzerdefinierten Funktionen. Zur Analyse dynamischer Systeme werden sowohl Werkzeuge für den Zeitbereich (wie Sprungantwort, Einschwingzeit) als auch für den Frequenzbereich (z.B. Frequenzgang) angeboten. Für den Entwurf von Regelsystemen können entweder klassische Methoden (wie PID-Regler) oder "moderne" Ansätze (wie Zustandsregler) eingesetzt werden. Die Berücksichtigung von Modellunsicherheiten (Robust Control) ist ebenso möglich wie die Formulierung adaptiver Regler.
Die Module "Digital Signal Processing" (DSP) und "System Identification" erlauben u.a. den Entwurf digitaler Filter, Frequenzanalysen von Meß- oder Simulationsergebnissen und die Bestimmung freier Modellparameter aus Messungen. Werkzeuge zur Modellreduktion werden im Modul "Model Reduction" angeboten.

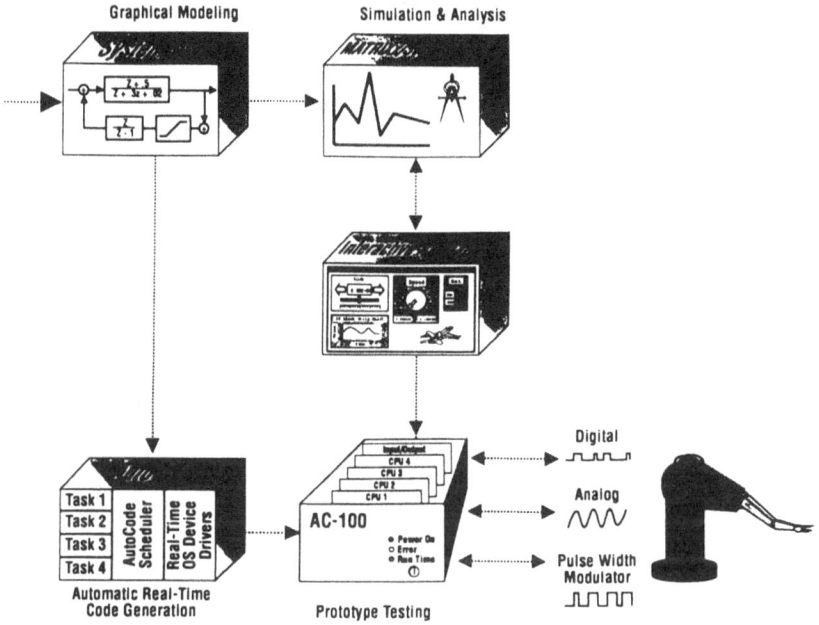

Bild 1: MATRIXx im Entwicklungsprozess von komplexen Regelsystemen

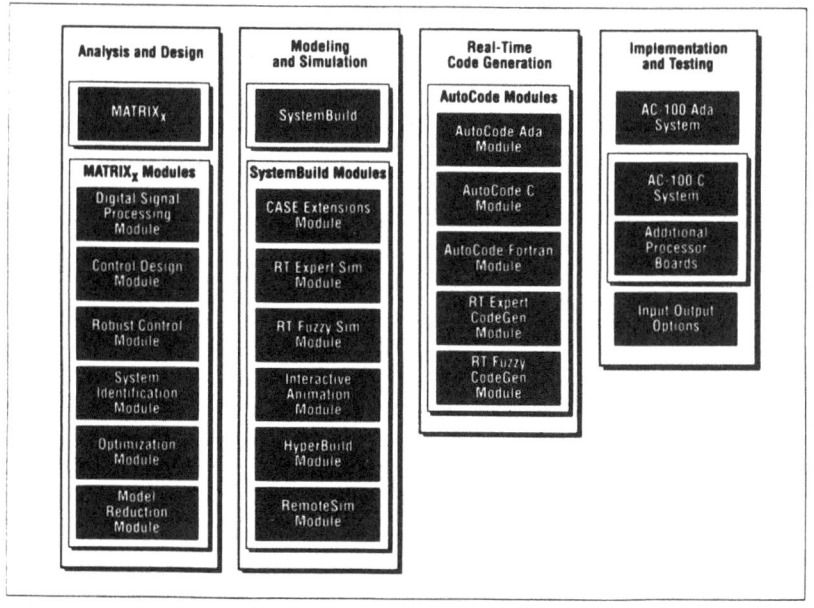

Bild 2: MATRIXx-Produktfamilie

2.2 Modellierung und Simulation

Innerhalb der MATRIXx Produktfamilie nimmt das Modul SystemBuild eine zentrale Stellung ein. In Form von Blockschaltbildern wird mit Hilfe des grafisch-interaktiven SystemBuild-Editors das dynamische System einschließlich Regler abgebildet. Aus einer vorhandenen Bibliothek können mehr als 70 Standardblöcke ausgewählt werden.
Während bei der ersten Version von SystemBuild klassische Übertragungsglieder im Vordergrund standen, ist in der Zwischenzeit die Bibliothek durch Blöcke erweitert worden, die die Modellierung von komplexen logischen Abläufen und wissensbasierenden Systemen ermöglicht. Zu nennen sind hier die Module "Case Extension" (Zustandsübergangsdiagramme und Data Stores), "RT/Expert" (Expertensystem für Echtzeitanwendungen) und das Modul "RT/Fuzzy".
Die Makrofähigkeit von SystemBuild erlaubt es, Teilsysteme in Form von sogenannten Superblöcken zu definieren. Diese Superblöcke können in einem Modell mehrfach verwendet werden und auch wiederum Teil eines weiteren Superblocks sein. Die Anzahl der Hierarchieebenen ist hierbei unbegrenzt.

Mit dem Modul "Interactive Animation" ist es möglich, simultan zu einer Simulationsrechnung Ein- und Ausgabegrößen grafisch darzustellen. So können z.B. Reglerparameter während der Simulation interaktiv modifiziert und das geänderte Systemverhalten beurteilt werden.

2.3 Codegenerierung

Die im Rahmen von SystemBuild modellierten Regelalgorithmen können mit Hilfe von "AutoCode" automatisch in echtzeitfähige Software (Sprachen: C, Fortran oder Ada) umgewandelt werden. Der Anwender muß lediglich einige hardwarespezifische Prozeduren an die gewählte Zielhardware anpassen. Mit "AutoCode" wird somit die Zeit zwischen Definition der Regelalgorithmen und erster Erprobung des realen Systems drastisch verkürzt. Weiterhin ist sichergestellt, daß Simulation und Reglerhardware stets die gleichen Algorithmen verwenden.

2.4 Implementierung und Prototypenversuch

Eine vollständige Integration von Reglerentwurf, Simulation, Implementierung und Prototypenversuch gestattet die Reglerhardware AC100. Ist der Regelalgorithmus in SystemBuild modelliert und simuliert worden, wird mit AutoCode die Umsetzung in echtzeitfähige Software vorgenommen, die Software kompiliert, mit speziellen Treiberroutinen gebunden, auf die Zielhardware geladen, und es kann unmittelbar danach das reale System getestet werden.

3. Das Modul RT/Fuzzy

Bei der Beschreibung von unscharfer Logik mit Hilfe des RT/Fuzzy-Moduls sind die folgenden Arbeitsschritte zu vollziehen (wobei die angegebene Reihenfolge nicht zwingend ist): Formulierung der Regeln, Deklaration der verwendeten Variablen (physikalisch oder unscharf) und Definition der Zugehörigkeitsfunktionen. Anschliessend ist zu bestimmen, nach welchen Methoden diese Regeln abgearbeitet werden sollen.

Eine Teilaufgabe innerhalb der ABS-Regellogik, nämlich die Ermittlung des Fahrbahnreibwertes, diene der Illustration der erforderlichen Arbeitsschritte. Als Eingangsgrößen seien der Bremsdruck und der Radschlupf gegeben. Mit Regeln der Art:
- falls der Bremsdruck niedrig ist und der Radschlupf groß, ist der Reibwert niedrig
- falls der Bremsdruck hoch ist und der Radschlupf klein, ist der Reibwert hoch

kann nun ein einfacher "Reibwertdetektor" erstellt werden. In RT/Fuzzy werden diese Regeln folgendermaßen formuliert:

    IF DRUCK IS NIEDRIG AND
      SCHLUPF IS HOCH
    THEN REIBWERT IS NIEDRIG

Nach der Eingabe der Regeln, sind die Variablen zu deklarieren und die Zugehörigkeitsfunktionen festzulegen. In dem Datensatz für den Bremsdruck

    DATA DRUCK
    TYPE CRISP
    RANGE [0 120]
    MEMBERS NIEDRIG GLOBAL
      MITTEL MAX(SIN(PI*X),0)**2
      HOCH [0, 0.2, 0.5, 1]

wurden beispielhaft verschiedene Arten der Beschreibung der Zugehörigkeitsfunktionen verwendet: Die Zugehörigkeitsfunktion "NIEDRIG" nimmt Bezug auf die globale Klasse NIEDRIG. MITTEL und HOCH werden hier lokal definiert und zwar MITTEL durch eine kontinuierliche Funktion und HOCH durch eine äquidistante Folge von Stützpunkten. Die Zugehörigkeitsfunktionen werden immer im Bereich 0 bis 1 formuliert. Die Zuordnung zu den physikalischen Größen erfolgt über die RANGE-Anweisung, hier Bremsdruck zwischen 0 und 120 bar. Die graphische Benutzerschnittstelle von RT/Fuzzy ist in Bild 3 dargestellt.

Die Behandlung der unscharfen Regeln erfolgt in RT/Fuzzy in den Schritten: Fuzzification, Connective Process, Implication, Defuzzification und Aggregation, wobei bis auf den ersten Schritt jeweils verschiedene Methoden zur Verfügung stehen, die noch durch benutzerdefinierte erweitert werden können.

Bild 4 zeigt den Berechnungsablauf für die Regel "falls der Bremsdruck niedrig ist und der Schlupf groß, ist der Reibwert niedrig". Im Schritt Fuzzification werden die physikalischen Größen, hier 50 % Schlupf und 40 bar Bremsdruck, in Fuzzy-Variablen umgewandelt, die den Zugehörigkeitsgrad zu den jeweiligen Klassen beschreiben.

Im Connective Process wird nun die Verküpfung der beiden Fuzzy-Variablen vorgenommen. Hierzu stehen zwei Methoden (MIN-MAX oder BAYESIAN) zur Auswahl, die durch eine benutzerdefinierte Methode ergänzt werden können.

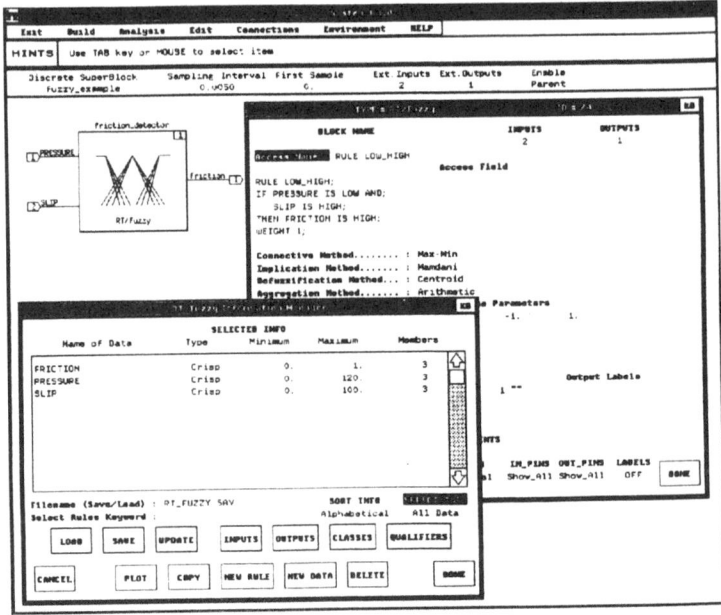

Bild 3: Benutzerschnittstelle von RT/Fuzzy

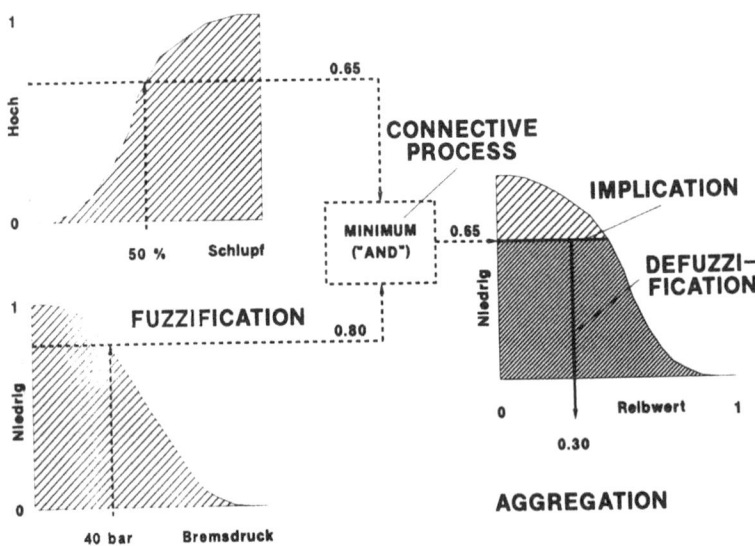

Bild 4: Berechnungsablauf von RT/Fuzzy

Abhängig vom Ergebnis des Connective Process ist nun die Zugehörigkeitsfunktion des Reibwertes im Schritt Implication zu ändern. Nach Mamdani wird die Funktion auf den Eingangswert, hier 0.65, begrenzt. Larsen skaliert die Funktion auf das Eingangssignal. Beide Lehrmeinungen sind in RT/Fuzzy implementiert und können durch eigene Ansätze erweitert werden.

Bei dem Schritt Defuzzification, der Umwandlung in eine physikalische Größe, besteht die Auswahl zwischen Schwerpunktbildung, Mittelwert der Maxima und selbst zu formulierenden Gesetzmäßigkeiten.

Der so beschriebene Berechnungsablauf wird nun für jede Regel vorgenommen. Der letzte Schritt, Aggregation, beschreibt dann, in welcher Weise z.B. fünf Regeln, die alle den Reibwert ermitteln, zu kombinieren sind. Hierzu werden verschiedene Methoden zur Mittelwertbildung angeboten, die wiederum durch eigene Ansätze erweitert werden können.

Das Verhalten von so erstellten Fuzzy-Regeln kann nun mit dem SystemBuild-Simulator entweder alleine oder in Verbindung mit anderen Regelelementen und Streckenbeschreibungen simuliert und optimiert werden. Weiterhin besteht die Möglichkeit, über vorhandene Linearisierungsalgorithmen Stabilitätsbetrachtungen durchzuführen.

Zeigt die Simulation schließlich das gewünschte Verhalten, kann mit Hilfe des AutoCode-Moduls echtzeitfähiger C- oder Ada-Code erzeugt werden. Je nach gewählter Option wird eine Optimierung hinsichtlich Speicherplatz oder Rechenzeit durchgeführt.

4. Anwendungsbeispiel

Zur Demonstration des Einsatzes der MATRIXx-Produktfamilie im Entwicklungsprozeß von komplexen Regelungsaufgaben wurde von Integrated Systems ein balancierender Aktenkoffer aufgebaut, Bild 5. Zwei gekoppelte Gewichte, die von einem Elektromotor angetrieben werden, halten das System im Gleichgewicht.

Die folgenden Sensoren erfassen die Bewegungszustände:
- ein inkrementaler Weggeber zur Erfassung der Position der Gewichte,
- ein Winkelsensor, der normalerweise zur Ausnivellierung von Wohnmobilen eingesetzt wird, und
- ein Kreiselsensor, der für Modellhubschrauber entwickelt wurde.

Im Unterschied zu dem akademischen Beispiel des invertierten Pendels besitzt das Demonstrationsobjekt eine Reihe von Eigenschaften, die typisch für industrielle Anwendungen sind:
- Haft- und Gleitreibung spielen eine wichtige Rolle.
- Der Verstellweg der Gewichte ist begrenzt.
- Es wurden preisgünstige Sensoren verwendet, mit dem Nachteil eingeschränkter Dynamik (Winkelsensor mit einer Eckfrequenz von 0.5 Hz) und hohem Rauschanteil (Kreiselsensor).

Bild 5: Anwendungsbeispiel

Zuerst wurden für dieses Problem die bekannten Methoden der Regelungstechnik eingesetzt und ein Zustandsregler entworfen, der aufgrund der komplexen Sensordynamik 14. Ordnung ist. Dem Zustandsregler nachgeschaltet sind Algorithmen, die eine Reibungskompensation durchführen. Mit dieser ersten Version konnten Störungen sehr gut ausgeregelt werden; es traten jedoch im ungestörten Zustand relativ große Schwingungen um die Gleichgewichtslage auf.
Diese Schwingungen konnten durch die Verwendung eines zweiten Satzes von Verstärkungsfaktoren reduziert werden. Wird eine größere Störung detektiert, schaltet die Regelung auf den ersten Satz von Verstärkungsfaktoren um, da andernfalls die Störung nicht ausgeregelt werden kann. Bild 6a zeigt schematisch den Kern des implementierten Reglers.

Alternativ zu dem rein "konventionellen" Regler wurde anschließend der Teil, der in Bild 6a eingerahmt ist, durch einen Fuzzy-Block ersetzt. Die Regelung erfolgt nun entsprechend Bild 6b: Der Zustandsbeobachter schätzt aus den Sensordaten die vier Bewegungszustände des Systems, aus denen im nachfolgenden Fuzzy-Block auf das erforderliche ideale Motormoment geschlossen wird. Wie beim rein konventionellen Regler wird anschließend noch eine Reibungskompensation durchgeführt. Simulation und Versuche am realen System haben gezeigt, daß mit insgesamt 10 Regeln ein befriedigendes Gesamtverhalten erzielt werden kann.

Ein Vorteil gegenüber dem rein "konventionellen" Ansatz ist darin zu sehen, daß nun kein hartes Schalten zwischen zwei Reglern erforderlich ist. Es ist jedoch anzumerken, daß nach wie vor ein Zustandsbeobachter implementiert ist. Zumindest bei diesem Beispiel wird es nicht sinnvoll sein, ausschließlich Fuzzy-Logic zu verwenden.

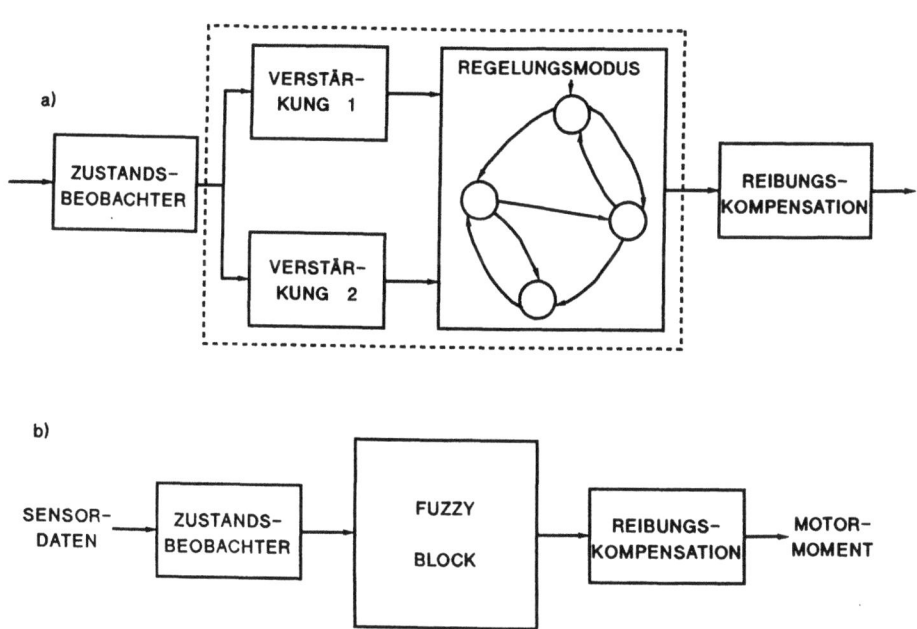

Bild 6: Regelungsansatz, a) rein konventionell, b) mit Fuzzy Logic

Drehzahlregelung eines Gleichstrommotors-konventionell oder mit Fuzzy-Logik?

Thomas Thiel

hema Elektronik GmbH
Röntgenstr. 31
W-7080 Aalen
Tel.: +7361 / 44031
FAX : +7361 / 44030

1 Einleitung

Der folgende Vortrag basiert auf einer Diplomarbeit, die in der Zeit zwischen dem 1. September 1991 und dem 29. Februar 1992 in Zusammenarbeit mit der Fachhochschule Aalen und der Firma hema in Aalen erstellt wurde.
Das Thema der Diplomarbeit lautete "Industrieapplikation mit Fuzzy-Logik". Ausgehend von einer einfachen Drehzahlregelung eines Gleichstrommotors, die mit konventionellen Methoden sehr gut beherrscht wird, soll untersucht werden, ob für die gleiche Aufgabe ein Fuzzy-Regler entworfen werden kann. Er sollte gleich gute oder bessere Eigenschaften wie der konventionelle Regler besitzen.
Welche Schwierigkeiten tauchen hierbei auf, worin unterscheiden sich konventioneller und Fuzzy-Regler?
Hauptziel war hierbei nicht die Optimierung beider Regler, sondern eine prinzipielle Betrachtung dieser Vorgehensweise. Der abschließende Vergleich beider Regler erfolgt daher unter allgemeinen Gesichtspunkten.

2 Übersicht

Im folgenden wird zunächst der prinzipielle Aufbau der Drehzahlregelung dargestellt. Anschließend wird die konventionelle Regelung kurz erläutert. Hieraus werden die Anforderungen an einen Fuzzy-Regler abgeleitet. Die beim Entwurf des Fuzzy-Reglers wichtigen Parameter werden anschließend beschrieben und bei der Beschreibung des prinzipiellen Vorgehens zum Entwurf eines Fuzzy-Reglers verdeutlicht. Abschließend werden konventioneller und Fuzzy-Regler anhand einiger Kriterien verglichen.

3 Regelung eines Gleichstrommotors mit dem PC

Beim als Regler arbeitenden PC handelt es sich hier um einen 386sx mit einer Taktfrequenz von 16 MHz. Die Software ist in C geschrieben. Der Fuzzy-Regler wurde in der FPL (Fuzzy Programming Language) von TOGAI beschrieben. Ein Durchlauf durch den TOGAI-Präprozessor liefert eine C-Funktion, die zum Hauptprogramm als externe Funktion gelinkt werden kann. Hierbei muß dann nur auf eine saubere Übergabe der Variablen geachtet werden. Der Fuzzy-Regler ist hier also rein softwaremäßig implementiert.

Die Ausgabe der Stellgröße erfolgt in Form eines pulsweitenmodulierten Signals mit einer Frequenz von 270 Hz an einem Pin der parallelen Schnittstelle.
Dieses Signal muß für den Motor in eine analoge Spannung umgewandelt werden. Der Motor ist für eine maximale Motorspannung von 12 V ausgelegt. Hierbei erreicht er eine Drehzahl von 4300 Umdrehungen/min.
Zur Erfassung der aktuellen Drehzahl besitzt der Motor einen integrierten Inkrementalgeber. Dieser hat zur Richtungserfassung zwei Kanäle, die um 90 Grad phasenverschoben sind. Für eine Drehzahlregelung reicht jedoch ein Kanal aus.
Abhängig von der Drehzahl hat das sinoïde Inkrementalgebersignal eine Frequenz von maximal 13 kHz. Zur Weiterverarbeitung mit dem PC muß es in ein Rechtecksignal konvertiert werden. Für den verwendeten PC muß die Frequenz anschließend durch den Faktor 4 geteilt werden um sicherzustellen, daß jeder Impuls vom Inkrementalgeber erfasst wird. Dieses niederfrequentere Signal wird dann über einen Pin der parallelen Schnittstelle eingelesen.
Die Hardware außerhalb des PCs befindet sich auf einer kleinen Platine, die direkt auf die parallele Schnittstelle aufgesteckt wird.
Abbildung 1 verdeutlicht die oben beschriebene Konfiguration.

4 Konventionelle (PID-) Regelung

Die prinzipielle Form eines PID-Algorithmus' ist in Abbildung 2 oben dargestellt. Die drei Anteile (P-, I- und D-Anteil) werden additiv überlagert. Sie sind durch die Faktoren $F_1...F_3$ gewichtbar. Durch Nullsetzen des entsprechenden Faktors kann so beispielsweise auch ein PI-Regler realisiert werden kann.
Die Ausgabe der Stellgröße ist durch die beiden folgenden Algorithmen möglich:

1. Positionsalgorithmus: hier wird jeweils der absolute Wert der Stellgröße ausgegeben

2. Geschwindigkeitsalgorithmus: hier wird die Änderung der Stellgröße zur letzten Stellgröße addiert, bzw. von ihr subtrahiert.

Abbildung 2 unten zeigt die beiden Algorithmen.

5 Anforderungen an einen Fuzzy-Regler

Welchen Anforderungen muß nun ein Fuzzy-Regler gerecht werden, um bei der Drehzahlregelung eines Gleichstrommotors mit konventioneller Regelungstechnik konkurrieren zu können?
Das System mit einem Fuzzy-Regler muß absolut stabil sein. Außerdem sollte der Fuzzy-Regler eine möglichst kurze Ausregeldauer besitzen. Die bei Störungen oder Führungsgrößenänderungen auftretenden Über-, bzw. Unterschwinger sollten

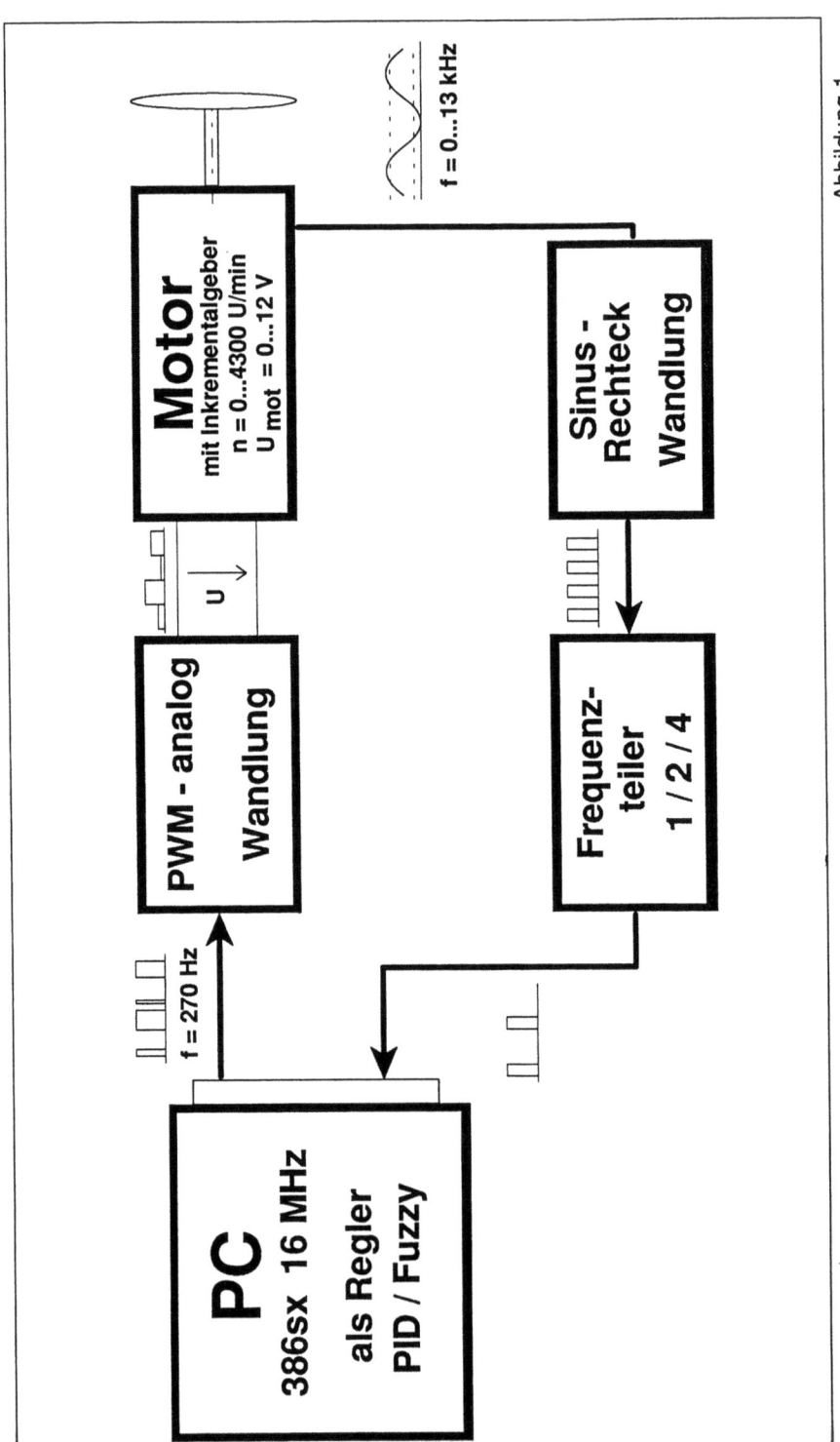

Abbildung 1

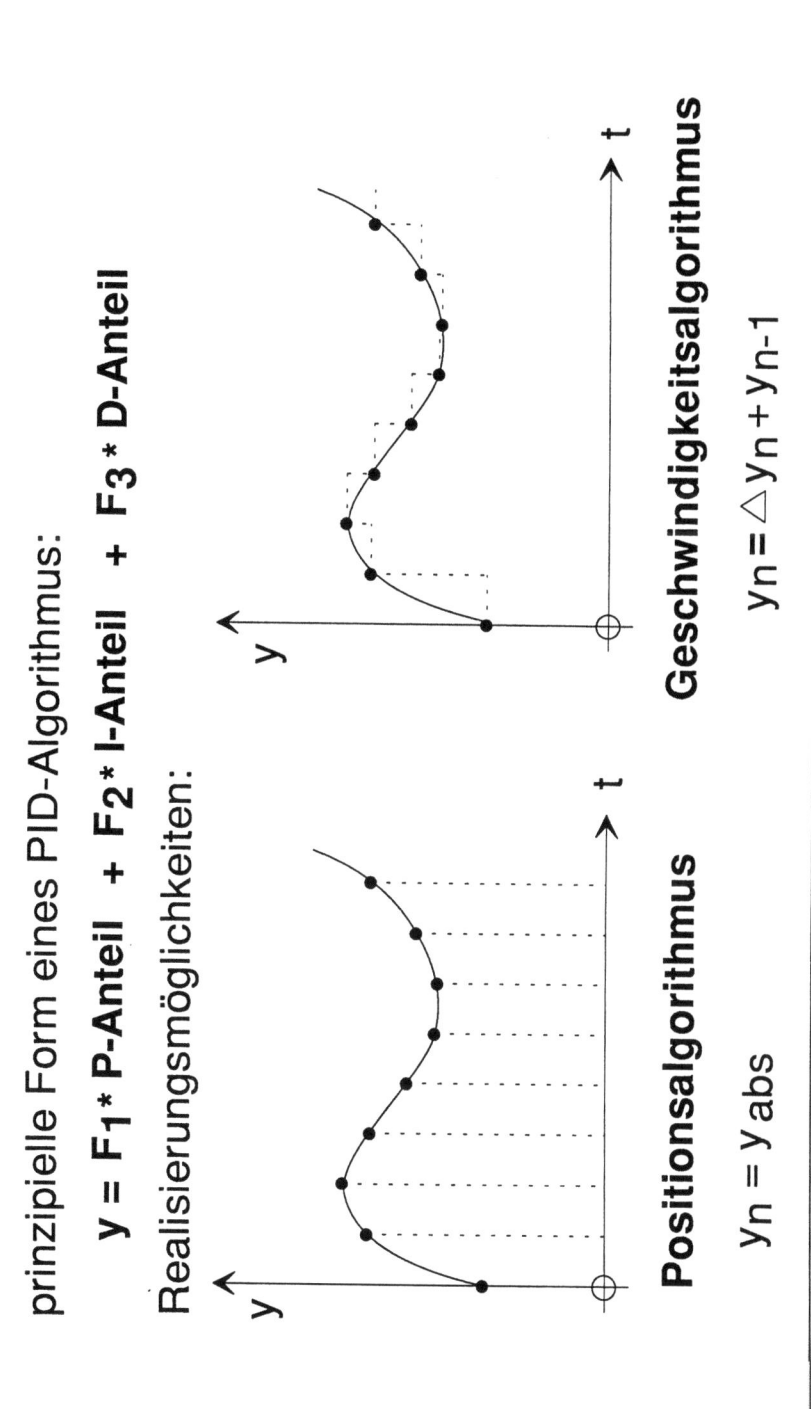

Abbildung 2

minimal sein. Ideal wäre eine Annäherung an den aperiodischen Grenzfall. Schließlich sollte der Entwicklungsaufwand zur Realisierung dieses Reglers angemessen sein.

6 Parameter beim Entwurf eines Fuzzy-Reglers

Beim Entwurf eines Fuzzy-Reglers nach o.g. Lastenheft sind folgende Parameter von Bedeutung:

1. Definition der Variablen: hier reichen für die Drehzahlregelung zwei Eingangs- (aktuelle Regeldifferenz und Änderung der Regeldifferenz) und eine Ausgangsvariable (Änderung der Stellgröße) aus.
2. Zugehörigkeitsfunktionen: die Anzahl der Terme (gerade/ungerade), ihre Lage (symmetrisch/unsymmetrisch) sowie die Form der Zugehörigkeitsfunktionen beeinflussen das Verhalten des Reglers. Für die Drehzahlregelung wurden pro Variable fünf Terme gewählt, deren Anordnung symmetrisch ist. Die Form der Zugehörigkeitsfunktionen ist dreieckig, bzw. dreieckig mit Schulter.
3. Regelbasis: hier sind die verwendeten Operatoren, die Aussage sowie die Anzahl der Regeln wichtig. Für die Drehzahlregelung wurde nur der UND-Operator verwendet, für eine stabile Regelbasis reichten 13 Regeln aus.
Liegt kein Bedienerwissen in Form von Wenn...Dann-Regeln vor, müssen die Regeln ermittelt werden. Dies kann auf folgende Art und Weise geschehen (siehe Abbildung 3 unten): ist beispielsweise die aktuelle Regeldifferenz Null UND die Änderung der Regeldifferenz Negativ, so muß die Stellgröße nicht geändert werden, da sie sich auf die Führungsgröße zubewegt; sie ist also Null. Auf diese Weise kann ein Regelsatz zusammengestellt werden.
Die Anzahl und die Aussagen der Regeln ist entscheidend für die Stabilität der Regelbasis.

7 Prinzipielles Vorgehen beim Entwurf eines Fuzzy-Reglers

Der Entwurf eines Fuzzy-Reglers ist iterativ (siehe Abbildung 4).
Zunächst müssen die Variablen definiert werden. Anschließend folgt eine grobe Festlegung der Zugehörigkeitsfunktionen. Falls Bedienerwissen in Form von Wenn...Dann-Regeln nicht existiert, müssen die Regeln ermittelt werden. Dies kann wie oben beschrieben geschehen. Dieser Regelsatz muß dann auf Stabilität hin überprüft werden. Dies kann entweder in der Simulation oder, bei der Drehzahlregelung, im Regelkreis geschehen. Stellt sich heraus, daß der Regelsatz nicht stabil ist, muß er optimiert werden. Dies geschieht, indem Regeln verändert, hinzugefügt oder weggelassen werden.
Anschließend erfolgt ein weiterer Test, etc. Der Regelsatz muß vor der Optimierung der Zugehörigkeitsfunktionen stabil sein, da diese nur eine Feinabstimmung darstellen. Eine nicht-stabile Regelbasis wird nicht durch optimale Zugehörigkeitsfunktionen stabil. Die Regelbasis sollte in möglichst extremen Zuständen getestet werden, da dies ein stabiles Verhalten im "Normalbetrieb" sicherstellt.
Ist die Stabilität der Regelbasis gewährleistet, kann das Reglerverhalten optimiert werden. Dies geschieht durch die Veränderung von Form und Lage der Zugehörig-

- **Definition der Variablen:**
  Eingangsvariablen: aktuelle Regeldifferenz
  Änderung der Regeldifferenz
  Ausgangsvariable: Änderung der Stellgröße

- **Zugehörigkeitsfunktionen:** Anzahl der Terme, Lage, Form

- **Regelbasis:** Operatoren, Aussage der Regeln,
  Anzahl der Regeln

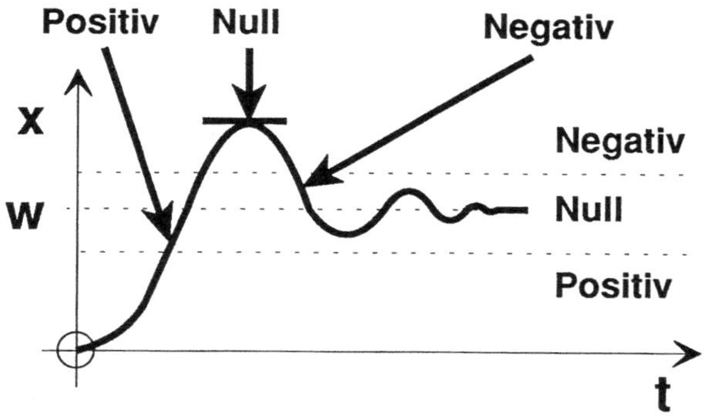

Beispiel für eine Regel:

WENN (aktuelle Regeldifferenz = Null)
UND (Änderung der Regeldifferenz = Negativ)
DANN (Änderung der Stellgröße = Null)

Abbildung 3

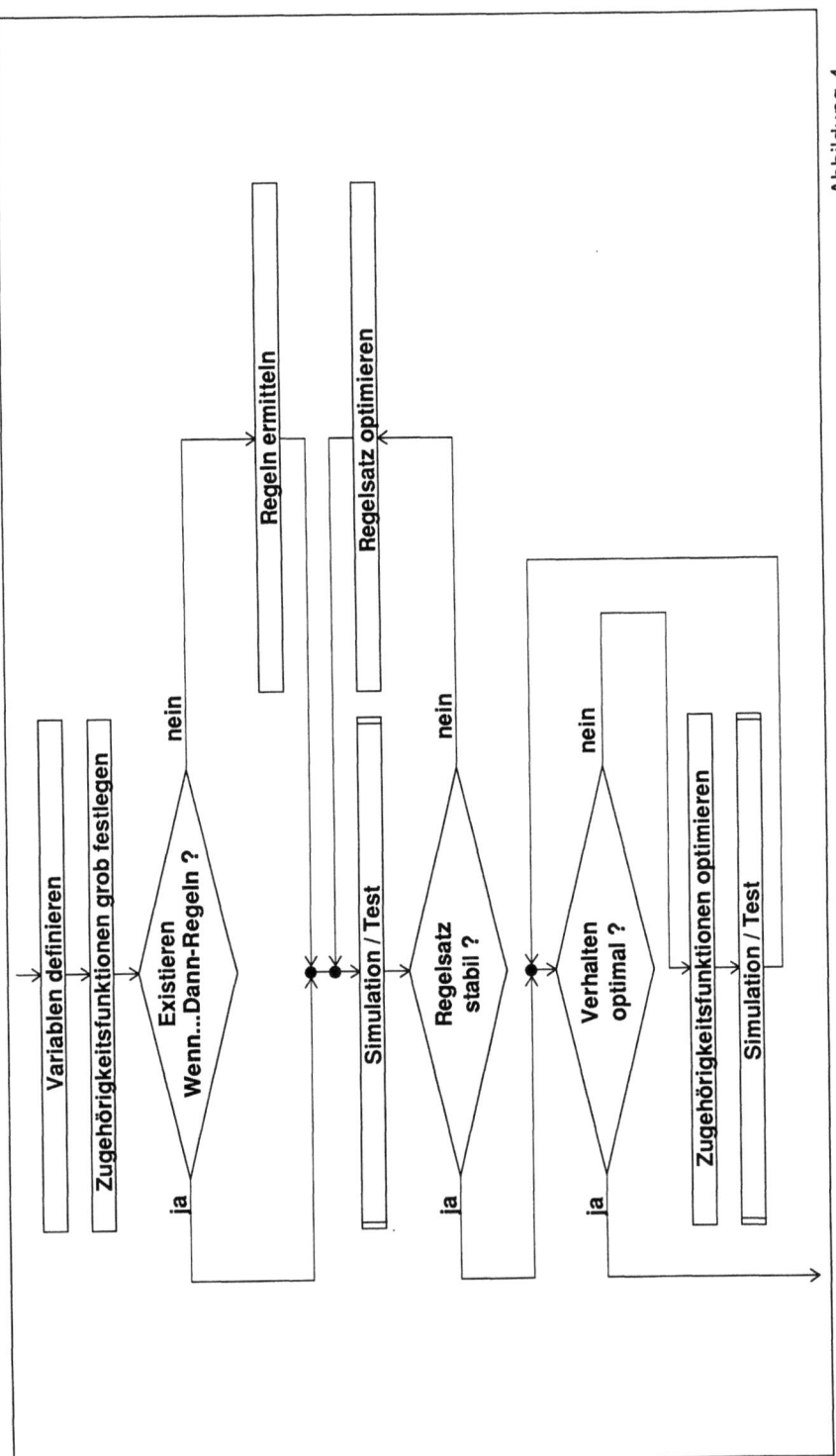

Abbildung 4

keitsfunktionen. Nach jeder Änderung wird in der Simulation, bzw. im Test überprüft, ob das Reglerverhalten als optimal zu bezeichnen ist.
Es empfiehlt sich, die Änderungen sorgfältig zu dokumentieren. Nur so ist es möglich, Änderungen, falls erforderlich, rückgängig zu machen.
Dieses geschilderte iterative Vorgehen resultiert aus dem Fehlen eines formalen Stabilitätskriteriums für Fuzzy-Regler.

8 Vergleich von konventionellem und Fuzzy-Regler

Konventioneller und Fuzzy-Regler sollen nun anhand verschiedener Kriterien verglichen werden.

1. Entwicklungsaufwand: verglichen mit einem konventionellen Regler ist der Entwicklungsaufwand für einen Fuzzy-Regler als hoch zu bezeichnen, wenn kein Bedienerwissen in Form von Wenn...Dann-Regeln vorhanden ist. Ist dies jedoch der Fall, reduziert sich der Aufwand für den Fuzzy-Regler stark. Nach der Stabilitätsprüfung müssen nur noch die Zugehörigkeitsfunktionen optimiert werden.

2. Aufwand zur Einstellung/Optimierung: die Einstellung eines konventionellen Reglers ist mit Erfahrung verbunden. Ist diese nicht vorhanden, steigt der Aufwand stark an. Beim Fuzzy-Regler ist bei vorhandener stabiler Regelbasis ein Hintrimmen auf eine bestimmte Reglercharakteristik relativ einfach.

3. Aufwand zur Einstellung des aperiodischen Grenzfalls: bei der Drehzahlregelung mit dem Fuzzy-Regler konnte der aperiodische Grenzfall sehr leicht realisiert werden.

4. Rechenaufwand für 1000 Stellgrößen: der Rechenaufwand bei einem Fuzzy-Regler hängt von der Komplexität der Regelbasis, der Anzahl der Variablen sowie der Anzahl der Terme pro Variable ab. Der Rechenaufwand hängt außerdem davon ab, ob die Berechnungen der Zugehörigkeiten während der Laufzeit durchgeführt, oder während des Präprozessorlaufs in Arrays abgelegt werden. Letzteres verringert natürlich den Rechenaufwand, bläht aber den Speicherbedarf sehr stark auf. Abbildung 5 unten zeigt den Rechenaufwand für 1000 Stellgrößen in Millisekunden in Abhängigkeit vom verwendeten Rechnertyp. Diese Zahlen sind aber nur in Relation zueinander zu betrachten. Es wird deutlich, daß der Rechenaufwand beim Fuzzy-Regler wesentlich größer ist. Hierbei ist aber zu berücksichtigen, daß der Fuzzy-Regler nur als Software implementiert ist. Eine Hardwarelösung würde das Verhältnis zugunsten des Fuzzy-Reglers wesentlich verbessern.

9 Zusammenfassung

Abbildung 6 zeigt eine Zusammenfassung. Als Wichtigstes bleibt festzuhalten, daß die Drehzahlregelung eines Gleichstrommotors mit einem Fuzzy-Regler prinzipiell möglich ist.
Beim Entwurf dieses Reglers haben sich folgende Vorteile herausgestellt:

Kriterium	konventionell	Fuzzy
Entwicklungsaufwand	niedrig	hoch
Aufwand zur Einstellung/Optimierung	hoch	niedrig
Aufwand zur Einstellung des aperiodischen Grenzfalls	hoch	niedrig
Rechenaufwand für 1000 Stellgrößen in ms		13 Regeln, 3 Variable je 5 Terme
		Arrays | Laufzeit
386sx 16 MHz	752,6	~30 000 | ~60 000
386sx 16 MHz + Coprozessor	49,1	1967 | 4170
386 20 MHz + Coprozessor	38,4	1489 | 3071
486 33 MHz	5,4	274,7 | 461,5

Abbildung 5

Fazit: Regelung eines Gleichstrommotors ist mit einem Fuzzy-Regler möglich

Vorteile	Nachteile
- bei Bedienerwissen in Form von Wenn...Dann-Regeln geringer Entwicklungsaufwand	- großer Rechenaufwand pro Stellgröße
- minimale Überschwinger leicht realisierbar (aperiodischer Grenzfall)	- großer Entwicklungsaufwand bei fehlender Regelbasis
- leichte Übertragbarkeit der Regelbasis auf andere Strecken	- oft "trial and error"-Vorgehen
	- fehlendes Stabilitätskriterium

Abbildung 6

1. Der Entwicklungsaufwand sinkt erheblich, wenn Bedienerwissen in Form von Wenn...Dann-Regeln vorhanden ist.

2. Der aperiodische Grenzfall war beim Fuzzy-Regler leicht zu realisieren; dies bedeutet eine Minimierung von Über- und Unterschwingern.

3. Existiert eine stabile Regelbasis, so ist sie auf andere Strecken mit ähnlicher Charakteristik leicht zu übertragen. Hierbei müssen nur noch die Zugehörigkeitsfunktionen der entsprechenden Strecke angepaßt werden.

Als Nachteile sind aufzuführen:

1. Der Rechenaufwand ist wesentlich größer als bei einem konventionellen Regler.

2. Existiert kein Bedienerwissen, d.h., die Regeln müssen einzeln ermittelt werden, steigt der Entwicklungsaufwand stark an. Dies hat den Grund, daß aufgrund des fehlenden Stabilitätskriteriums ein "trial and error"-Vorgehen notwendig wird.

Abschließend bleibt noch zu sagen, daß vermutlich nicht jede regelungstechnische Anwendung durch den Einsatz von Fuzzy-Logik optimiert werden kann. Am Beispiel des Gleichstrommotors wurde gezeigt, daß hier konventionelle Regelungstechnik von Vorteil ist. Die Stärken von Fuzzy-Reglern sind bei nichtlinearen Strecken, komplexen Systemen und bei Strecken mit Totzeiten deutlich zu erkennen.

Der Einsatz von Fuzzy-Prozessoren zur Klassifizierung und Analyse mechanischer Systeme

Vortrag Klaus Stieger/Firma IDS

1. Überblick

Zur Untersuchung mechanischer Systeme, aber auch in Bereichen wie Produktsicherung und Fertigungsüberwachung, ist die akustische Analyse mittlerweile eine wohletablierte Technik. Während Erfassung und Darstellung der Daten im allgemeinen heute keine Probleme mehr bereiten, erfordert die Interpretation und Auswertung der Messungen meist ein hohes Maß an Expertenwissen und Erfahrung, um Fehlaussagen zu vermeiden.

Mit der inzwischen drastisch angewachsenen Verarbeitungskapazität moderner Signalprozessoren und der damit verbundenen einfachen Möglichkeit der online Berechnung von FFT und Cepstrum haben sich zwar für die akustische Signalanalyse neue Perspektiven eröffnet, jedoch ist eine effiziente Einbringung von Expertenwissen bei der Etablierung von Kriterien zur automatischen Auswertung noch immer recht mühevoll.

Um diesbezüglich eine spürbare Vereinfachung zu schaffen und die Verläßlichkeit der Aussage auch bei komplexen Anwendungen zu optimieren, wurde ein Analysesystem entwickelt, welches auf der Basis von Fuzzy-Logic arbeitet.
Dadurch wird nicht nur die Implementierung der Bewertungsalgorithmen wesentlich vereinfacht und erweitert, sondern auch eine hohe Flexibilität bei der Anpassung an sich ändernde Aufgabenstellungen gewonnen.

Das nachfolgend beschriebene System dient der Erfassung und Analyse von akustischen Signalen im Audiofrequenzbereich. Alle DSP Aufgaben werden von Transputern übernommen. Die Klassifizierung bzw. Signalanalyse erfolgt durch einen speziellen Fuzzy-Prozessor.
Es kann wahlweise entweder der Zeitverlauf, das Spektrum oder aber auch das Cepstrum des Meßsignals bewertet werden.

Man erhält somit einen außerordentlich flexiblen, für die unterschiedlichsten Meßaufgaben vom Anwender leicht zu programmierenden Analyzer, der einer vielfältigen und wechselnden Aufgabenstellung im Bereich der Signalanalyse gerecht wird.

An Hand eines Beispieles, nämlich der Klassifizierung von Tennisbällen auf Grund ihres Aufprallgeräusches, wird vorgeführt, wie das Gerät an eine spezielle Aufgabe adaptiert werden kann.

2. Technisches Konzept

2.1 Prinzipien

Ziel der akustischen Analyse ist es, durch Erfassung und Bewertung der einem Vorgang oder Prozeß immanenten charakteristischen Geräuschemissionen bzw. Vibrationen, Information über Betriebszustände, aktuelle Prozeßparameter oder Funktionsmechanismen abzuleiten.

In aller Regel handelt es sich dabei um komplexere Schwingungsformen, so daß eine Analyse im Zeitbereich meist keine ausreichenden Rückschlüsse zuläßt. Hier bietet sich die Berechnung des Spektrums oder des Cepstrums an, um die Signalcharkteristika besser herauszuarbeiten und so einer Klassifizierung und Bewertung zugänglich zu machen.

Aufgabe automatischer Analysesysteme ist es nun, an Hand dieser typischen Merkmale Zustände, Betriebsparameter oder Klassenzugehörigkeit der vorliegenden Prozesse zu beurteilen und gegebenenfalls sogar, davon abgeleitet, korrigierend in den Prozeß einzugreifen. Das was einem menschlichen Experten durch Beurteilung der vorliegenden Signalcharakteristika im Allgemeinen gelingt, bereitete bei der Implementierung auf Digitalrechnern bislang teilweise erhebliche Schwierigkeiten. Dies nicht zuletzt deshalb, weil eine algorithmenmäßige Formulierung mit "klassischen" Verfahren nicht nur großen Aufwand darstellte, sondern auch die Einbeziehung der statistischen Schwankungen, wie sie bei realen Prozessen stets auftreten, sich als schwierig gestaltete. Besonders kommt dieser Umstand dann zum Tragen, wenn es erforderlich wird, z.B. wegen geänderter Prozeßparameter, eine Neuanpassung der Bewertungskriterien vorzunehmen.

Um das Expertenwissen, d.h. die Kenntnisse über die Eigenschaften des untersuchten Systems sowie die daraus abgeleiteten Kriterien zur Signalklassifizierung besonders effizient einbringen zu können, wurde bei dem im folgenden beschriebenen Analyzer Fuzzy-Logic eingesetzt. Dieses Verfahren ermöglicht neben der Formulierung der Bewertungs-Algorithmen in einer der menschlichen Ausdrucks- und Denkweise nahekommenden Syntax, auch Signale mit großer statistischer Interpretationsbreite zuverlässiger als mit herkömmlichen Methoden zu klassifizieren.

Prinzipiell läßt sich ein Meßvorgang mit diesem Gerät in folgende vier Phasen einteilen:

- Erfassung des Signals bzw. der Signale (Mehrkanalversion) im Zeitbereich und grafische Darstellung

- On-line Berechnung der Fouriertransformierten oder wahlweise des Cepstrums, ebenfalls mit grafischer Ergebnisdarstellung

- Bewertung der so erhaltenen Signale in Echtzeit mittels Fuzzy-Logic

- Ergebnisausgabe in einer vom Anwender wählbaren, der Problemstellung adäquaten Form.

Vom Standpunkt des Benutzers aus ist es besonders wichtig, daß jeder dieser Schritte durch flexible, anwenderfreundliche Software unterstützt wird, da nur so ein breites Spektrum von Aplikationen abgedeckt werden kann.

3. Realisierung und Aufbau

3.1 Hardware

Realisiert wurde das Analysesystem unter Einsatz modernster Rechnertechnologie. Transputer dienen zur Datenerfassung und zur Abarbeitung der DSP-Tasks, wie FFT oder Cepstrum-Berechnung. Ein spezieller Fuzzy-Prozessor, der Togai FC110, übernimmt die Auswertung der Bewertungsregeln und die Ermittlung der Ergebnisse. Die Struktur des Gesamtsystems ist aus Abb. 3.1-1 ersichtlich.

Das Gerät besteht aus der Analyzer Unit, die über einen Bus-Extender an den Host-PC angekoppelt ist. Lediglich ein 8 Bit Steckplatz im Host ist zur Anbindung erforderlich. Auf einer in der Analyzer Unit integrierten SCSI Platte wird nicht nur alle gerätespezifische Betriebs- und Anwendersoftware gehalten, sondern auch Ergebnis-Files und Data-Records können hier abgelegt werden. Die weiteren Funktionseinheiten des Gerätes sind das Analog Input Modul, das DSP Modul, sowie das Fuzzy-Prozessor Modul.

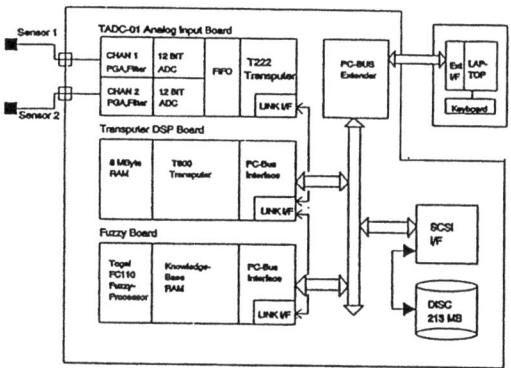

Abb. 3.1-1 Blockdiagramm des Analyzers

Die gesamte Analyzer Unit ist in Transputer Technologie aufgebaut, und die einzelnen Module sind durch Transputer-Links untereinander verbunden. Die Kommunikation zwischen dem Transputernetz und dem Host-PC erfolgt über eine spezielle Interface-Karte, die das serielle Linkprotokoll auf den ISA-Bus umsetzt. Bei der seriellen Kommunikation via Link werden effektive Datenraten von etwa 1.4 MByte/sec erzielt. Durch konsequente Anwendung der Transputer- und Link-Philosophie, ist der Analyzer sowohl bezüglich der Kanalzahl, als auch der DSP-Performance in fast beliebigen Grenzen erweiterbar.

Das Analog Input Modul umfaßt zwei unabhängige Eingangskanäle zum direkten Anschluß unterschiedlicher Sensoren wie z.B. Meßmikrofone oder piezoresistive Beschleunigungsaufnehmer. Jeder Kanal besitzt als Eingangsstufe einen Instrumentierungsverstärker mit softwaremäßig programmierbarem Verstärkungsfaktor von $V=1$ bis $V=1000$. Es folgt jeweils ein Tiefpaßfilter vierter Ordnung mit Butterworth-Charakteristik und ein bezüglich THD optimierter Sampling ADC mit 12 Bit Auflösung. Über FIFOS werden die AD-Wandler beider Kanäle an einen T222 Transputer angekoppelt, der die Datensammlung und Einspeisung in das DSP Transputernetz via Link übernimmt.

Die Abtastung beider Kanäle erfolgt mit simultan S&H, wobei Abtastraten von 1.25 kHz/Kanal bis 125kHz/Kanal softwaremäßig gesetzt werden können. Ein Trigger-Eingang ermöglicht die Synchronisation von DSP-Prozessen auf externe Ereignisse.

Darüber hinaus stellt das Analog Input Modul Funktionen wie Eingangskalibrierung, DC-Offsetshift, sowie die Sensorspeisung für Geber mit resistiver Vollbrücke zur Verfügung.

Das DSP Modul besteht in seiner Grundausbaustufe aus einem T800 Transputer mit integrierter FPU und 4Mbyte externem, schnellen RAM. Mit dieser Hardware läßt sich eine 1024 Punkte komplexe FFT in ca. 68 ms berechnen. Wird eine höhere

Rechenleistung benötigt, so erlaubt das Konzept die Aufteilung der Tasks auf mehrere Transputer, oder aber die Einbindung 'echter' Signalprozessoren sowohl hard- als auch softwaremäßig über Link.

Der Fuzzy-Prozessor, ein Togai FC110, wird ebenfalls über einen Transputer-Link an den Datenfluß angebunden. Der FC110 wurde speziell für die effiziente Abarbeitung von Fuzzy-Algorithmen designed, do daß durch den Einsatz dieser Hardware eine Durchführung der Klassifizierungsprozesse im ms Bereich möglich wurde.
Der 128kByte große, externe Speicher für Programmcode und Knowledge-Base wurde als RAM ausgeführt, das über den PC-Bus geladen werden kann. Damit erhält man die Möglichkeit, Fuzzy-Rules und Membership-Functions per Software direkt zu modifizieren.

3.2 Software

Es wurde eine grafische Benutzeroberfläche unter MS-Windows 3.0 implementiert. Über einen grafikfähigen Link-Server wird die DSP Applikation, die auf dem Transputernetz abläuft, an die Oberfläche angebunden. Die Fuzzy Applikation ist kommunikationsmäßig ebenfalls Bestandteil des Transputernetzes. Zur Generierung der Membership-Functions und der Rule-Base dient ein spezieller Editor, der eine recht komfortable Umsetzung des Expertenwissens in Fuzzy-Regeln ermöglicht. Ein Fuzzy-Toolset bestehend aus Compiler, Linker und Loader für die FC110 Zielhardware ist ebenfalls Bestandteil der Software.
Im Grundausbau des Gerätes wurde ein Satz Basisbefehle implementiert, die je nach Applikationsanforderung und Hardwareausbau kundenspezifisch erweitert werden können (siehe auch Kap. 4). Diese Befehle sind:

DISPLAY
Hier werden die Darstellungsarten (Zeitbereich, Betrag und Phase des Spektrums, Histogrammdarstellung, Ergebnisfenster usw.) gewählt. Außerdem dient dieser Menüpunkt zur Einstellung der FFT-Parameter.

PRINT
Durch Anwählen dieses Menüpunktes wird eine Kopie des aktuellen Grafikfensters im HP-GL Format erstellt. Zu Dokumentationszwecken kann ein beliebiger Text eingegeben werden

TRIGGER
Unter diesem Punkt lassen sich die Triggerarten bei der Signalacquisition festlegen. Es gibt die Modi Manual, Continuous, Single-Shot und Auto. Im Auto Mode wird, wenn das Signal eine vorher eingestellte Triggerschwelle überschreitet, automatisch eine der eingestellten FFT-Länge entsprechende Anzahl von Meßpunkten acquiriert. Im Continuous Mode erfolgt dies unabhängig vom Verlauf des Eingangssignales. In beiden Fällen ist es jedoch möglich, die FFT auf ein externes Triggerereignis zu synchronisieren.

WINDOW
In Abhängigkeit vom Signalverlauf kann der Anwender hier zwischen den Fensterformen Flat, Hanning und 3-Term Blackman-Harris wählen. Für die Analyse von transienten Signalen, wie im Beispiel aus Kap.4, muß ein Flat Window gewählt werden.

FUZZY
Im Fuzzy Menü werden die Tools wie Knowledge-Base Editor, Loader oder Compiler aufgerufen. Eine On-line Änderung der Membership-Functions ist möglich.

4. Systemadaption

Die Adaption des Analyzers an eine konkrete Problemstellung wird im folgenden an einem Anwendungsbeispiel erläutert. Der Schwerpunkt liegt dabei auf der Vorgehensweise, d.h. wie mit Systemunterstützung die Ableitung der Fuzzy-Rules und der Membership-Functions erfolgt. Die Aufgabenstellung lautet:

"Klassifizierung von Tennisbällen anhand ihres Aufprallgeräusches"

Bereits vorab kann man als Fazit folgende Aussagen festhalten: Nach Einführung einer Lernphase für die Erstellung der Membership-Functions arbeitet das System mit nahezu 100-prozentiger Trefferquote. Die eigentlichen Probleme lagen mehr in den physikalischen Randbedingungen als beim Einsatz der Fuzzy-Logic. Vielmehr erwies sich Fuzzy-Logic als sehr hilfreiches und flexibles Werkzeug, um die Klassifizierungsaufgabe einfach und effizient zu lösen, obwohl die akustischen Signale die ein und derselbe Ball beim Aufprall erzeugte unerwartet breit streuten.

4.1 Bedieneroberfläche

Wie in Kapitel 3.2 vorgestellt, arbeitet das System mit einer graphischen Oberfläche, welche unter MS-Windows abläuft. Standardmäßig erscheint der in Bild 4.1-1 dargestellte Bildschirm. Im unteren Teil wird das Zeitsignal und im oberen Teil das Amplitudenspektrum des Zeitsignals dargestellt.

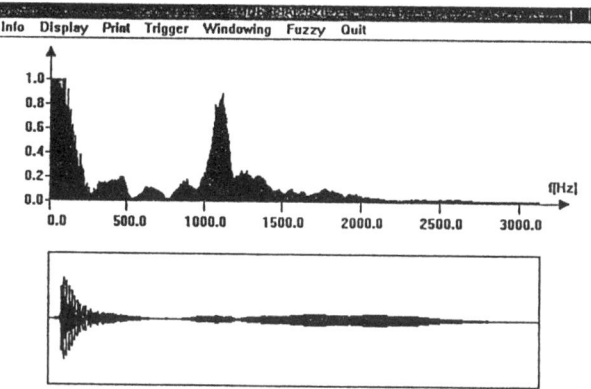

Bild 4.1-1: Standard Bildschirm

4.2 Ableitung der Fuzzy-Rules und der Membership-Functions

Nach einigen Würfen verschiedener Tennisbälle wurde sehr schnell deutlich, daß die gravierensten Unterschiede im Amplitudenspektrum der Bälle im Bereich zwischen 800 und 1500 Hz auftraten. Im Zeitbereich waren selbst bei genauester Beobachtung keinerlei Unterschiede erkennen. Um nun zu linguistischen Variablen der Art "sehr hoch", "hoch", "mittel", "niedrig" und "sehr niedrig", wie sie für Fuzzy-Logic benötigt werden, zu kommen, kann man über den Menüpunkt DISPLAY an Stelle der Zeitfunktion ein Balkendiagramm darstellen (Bild 4.2-1). In unserem konkreten Beispiel werden darin die Spektralanteile über den Bereich von 800 bis 1500 Hz in 10 äquisistanten Abschnitten aufsummiert und normiert dargestellt.

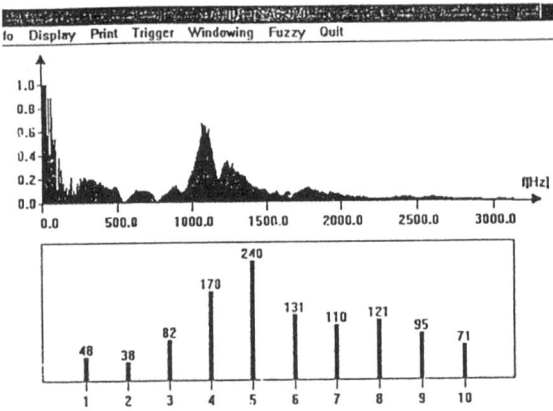

Bild 4.2-1: Darstellung von Amplitudenspektrum und Balkendiagramm

Ein erster Ansatz war nun, Membership-Functions in der bekannten Form überlappender Dreiecke zu definieren:

        0 - 75   "very low"
     25 - 125   "low"
     75 - 175   "middle"
    125 - 225   "high"
    175 - 250   "very high".

Als nächster Schritt können für jeden Ball Fuzzy Rules der Art

```
RULE Ball1_rule
    if   (H1 is very_low or H1 is low)
    and (H2 is low or H2 is middle)
    and (H3 is very_high)
    and (H4 is high or H4 is very_high)
    and (H5 is high or H5 is very_high)
    and (H6 is high or H6 is very_high)
    and (H7 is middle or H7 is High)
    and (H8 is low or H8 is middle)
    and (H9 is low or H9 is middle)
    and (H10 is low or H10 is middle)
        then Ball is Ball1_member
END
```

aufgestellt werden.

Danach bleiben noch folgende Fragen zu klären:

- Soll man eine Output Membership-Function für alle Bälle gemeinsam, oder soll man eine Function pro Ball definieren ?
- Wie sehen die Membership-Functions für die Ausgangsvariablen aus?
- Soll Defuzzifiziert werden, oder soll man besser mit den Alpha-Werten (Ergebnis nach Auswertung der Left Hand Side einer Regel) arbeiten?

Die Antwort auf diese Fragen lautet: Bei Klassifizierungsaufgaben ist ohne Defuzzifizierung zu arbeiten. Eine Begründung dafür gibt folgende Überlegung:

Verwendet man z.B. für jeden Ball eine Output Membership-Function mit der Form eines Dreiecks, so kann der Fall eintreten, daß zwei Regeln feuern, deren Output Membership-Functions nicht nebeneinander liegen. Eine Defuzzifizierung nach der Centroid-Methode liefert dann einen Wert (Schwerpunkt der beiden Flächen) der zwischen den beiden Flächen liegt, so daß als Ergebnis ein ganz anderer Ball erkannt wird.

Der beschriebene Ansatz führte jedoch noch nicht zu einer zufriedenstellenden Trefferquote bei der Erkennung der Bälle. Nach einer Analyse der Ergebnisse wurden als Modifikationen je Histogramm (H1,...,H10) Membership-Functions pro Ball definiert. Der Aufbau der Regeln ist nun wie folgt:

```
RULE Ball1_rule
    if  (H1 is Ball1_member)
    and (H2 is Ball1_member)
    and (H3 is Ball1_member)
    and (H4 is Ball1_member)
    and (H5 is Ball1_member)
    and (H6 is Ball1_member)
    and (H7 is Ball1_member)
    and (H8 is Ball1_member)
    and (H9 is Ball1_member)
    and (H10 is Ball1_meber)
    then Ball1 is ANY
END
```

Anstatt der Dreiecksfunktionen, wie man sie in den meisten Veröffentlichungen als Membership-Function vorfindet, wurden Gaußfunktionen verwendet. Die Bestimmung von Mittelwert und Varianz pro Histogramm und Ball, wurde mit Hilfe eines kleinen Zusatzprogrammes (Lernprogramm) für eine statistisch repräsentative Anzahl von Wurfversuchen bestimmt. Damit ergaben sich folgende Membership-Functions:

```
VAR H1
TYPE unsigned byte
    MEMBER Ball1_member
        EQUATION (EXP(((H1-60)/16)^2)/(-2))
    END
    MEMBER Ball2_member
        EQUATION (EXP(((H1-114)/19)^2)/(-2))
    END
    MEMBER Ball3_member
        EQUATION (EXP(((H1-62)/11)^2)/(-2))
    END
    MEMBER Ball4_member
        EQUATION (EXP(((H1-41)/14)^2)/(-2))
    END
END
```

Nach diesen Modifikationen war die Trefferquote nahezu 100 Prozent. Fuzzy-Logic erwies sich als flexibles, leicht handhabares Werkzeug. Die Änderungen der Knowledge-Base (Membership-Functions und Fuzzy-Rules) dauerte keine 3 Stunden. Die meiste Arbeit verbarg sich in der Erarbeitung der Selektionskriterien, d.h. in der Entwicklung eines "Expertenwissens" (starke Temperaturabhängigkeit und die unregelmäßige Oberfläche der Tennisbälle, hatten eine breite Signalstreuung zur Folge). Die fertig aufbereitete Darstellung der Klassifikationsergebnisse ist in Bild 4.2-2 zu sehen.

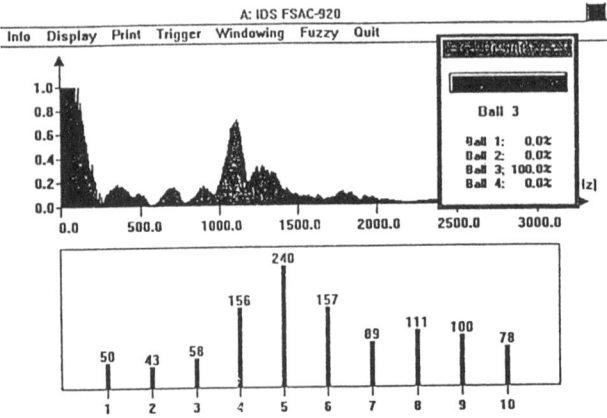

Bild 4.2-2 Ergebnisdarstellung

Die Bearbeitungszeit für einen Klassiervorgang, d.h. die Zeit vom Wurf des Balls bis zur Ausgabe der Ergebnisse, berträgt bei der kleinsten Ausbaustufe des Analyzers etwa 2 sec. Dabei entfällt der größte Anteil auf die grafische Bildschirmausgabe, die Rechenergebnisse selbst liegen zur Weiterverarbeitung nach etwa 700ms vor. Durch Hardwareausbau läßt sich dieser Wert um einen Faktor von etwa 10 bis 20 reduzieren.

5. Zusammenfassung und Ausblick

Im Rahmen der beschriebenen Systemimplementierung konnte gezeigt werden, daß Fuzzy-Algorithmen in Verbindung mit DSP sehr gut zur Mustererkennung und Klassifizierung geeignet sind. Eine besondere Bedeutung kommt dabei für viele Anwendungen der Echtzeitfähigkeit zu.
Durch die sowohl hard- als auch softwaremäßige Einbindung von DSP und Fuzzy-Logic in ein Transputernetz, wird auf unkomplizierte Weise die Realisierung von problemorientierten Topologien und Ausbaustufen möglich. Dabei stellen die Transputer mit ihrer schnellen Anbindung an das Host-System die Kommunikations-Infrastruktur, während an den einzelnen Netzknoten neben den Transputern selbst dedizierte Hardware wie Signalprozessoren oder Fuzzy-Prozessoren angebunden werden kann. Ein wesentlicher Pluspunkt ist dabei, daß das gesamte Netz, einschließlich der speziellen Prozessoren, zentral gebootet wird. Dadurch bleibt die Software-Konfiguration trotz der Kombination unterschiedlicher Prozessoren überschaubar, und eine dynamische Rekonfiguration des Gesamtsystems, z.B. für sich ändernde Aufgabenstellungen, wird möglich.
Mit einem jeweils problemorientierten Systemausbau kann somit den Belangen unterschiedlichster Applikationen entsprochen werden, sowohl bezüglich der I/O Kapazität als auch hinsichtlich der Rechenleistung.

Die Realisierung einer Fuzzy Logik Hardware Implementation am Beispiel des NeuraLogix Fuzzy MicroControllers

Oliver Breiden
U.S. Marketing Operations
UNITRONIC GmbH
Mündelheimer Weg 9
4000 Düsseldorf 30

1 Die Hauptstärken von Fuzzy Logik Applikationen

In den kommenden Jahren ist mit einem starken Anstieg von Fuzzy Logik Applikationen zu Rechnen. Insbensondere in den Bereichen Konsumgüter- und Unterhaltungselektronik, Alarm- und Sicherheitstechnik ist mit Zuwachsraten zu erwarten. In der Prozeßsteuerung bieten Fuzzy-Systeme oftmals eine sinnvolle Alternative zu den bisher verwendeten konventionellen Mikroprozessor-Steuerungen. Geht es darum komplexe und nichtlinear verlaufende Funktionen zu beschreiben, so ist es unter der Zuhilfenahme von Fuzzy-Logik Entwicklungstools möglich, die Entwicklungszeiten drastisch verkürzen. Der sich daraus ergebende wirtschaftliche Vorteil der beschleunigten Marktreife ("time-to-market") kann in der Regel zu einer erheblichen Kostenreduzierung beitragen. Verkürzte Entwicklungszeiten, Kosteneinsparungen in der Entwicklungphase und ein attraktives Preis-Leistungsverhältnis stellen die drei herausragenden Hauptmerkmale der Fuzzy-Logik dar.

2 Grundlagen einer Fuzzy Hardware Lösung

Will man die Grundlagen der Fuzzy Logik in eine Hardware Lösung implementieren, so gilt es eine Vielzahl von möglichen Kriterien zu beachten. In der Regel handelt es sich hierbei um in entgegengesetzte Richtungen strebende Lösungsmöglichkeiten. Hierzu ein Beispiel: ein hohes Maß an Flexibilität erfordert ein ebenfalls hohes Maß an Komplexität. Ist man also darum bemüht, das gesamte Feld der möglichen Fuzzy Applikationen mit einem Ansatz abzudecken, so wird man unwillkürlich Gefahr laufen, nicht kostenrentabel Entwickeln zu können. In ähnlichen Maße wie Flexibilität und Komplexität in direktem Gegensatz zueinander stehen, muß man Kriterien wie beispielweise Komplexität und Geschwindigkeit, Komplexität und Preis, Leistungsfähigkeit und Anwenderfreundlichkeit, sowie Leistungsfähigkeit und Leistungsbedarf gegeneinander Abwiegen. Das Spektrum der möglichen Hardware Implementierungen, und dies gilt im übrigen auch für Fuzzy-Compiler und andere Software-Tools, reicht somit vom relativ einfachen, low-cost Chip mit schnellem Datensurchsatz und limitierten Modifikationsmöglichkeiten, bis hin zum komplexen, high-end Expertensystem, welches flächendeckend das gesamte Feld der möglichen Fuzzy Applikationen erfasst.

2.1. Zielsetzung des NeuraLogix Fuzzy MicroControllers

Bei der Entwicklung des NeuraLogix NLX230 Fuzzy MicroControllers galt es die oben genannten Kriterienen in einem möglichst idealen Maße zu verknüpfen. Die folgenden Bedingungen waren bei der Entwicklung der NeuraLogix Chip Philosopie von Bedeutung:

- **Stand-alone Chip Lösung**, d.h. der Chip soll nach der Entwicklung von der Entwicklungsoberfläche abgekoppelt werden und autark arbeiten
- **Low-Cost**, d.h. der Chip soll in der Preislage mit konventionellen Microprozessoren konkurieren und insbesondere im High-Volume Bereich (bei großen Stückzahlen) eine attracktive Alternative bieten
- **Hoher Datendurchsatz**
- **Einfache Handhabung**, insbesondere im Bereich der Entwicklungsoberfläche
- Größtmögliche **Applikationsabdeckung**

In der Zielsetzung wurde bewußt darauf versichtet alle theoetisch definierbaren Fuzzy-Logik Problemstellungen abzudecken. Vielmehr galt es, eine möglichst weitspannende und kostenattraktive alternative zu konventionellen Micro-Controllern zu entwickeln. Der Hauptentwicklungsschwerpunkt lag in der Adaption der Fuzzy-Logik Philosophie für eine in größtem Maße applikationsdeckende Hardware Implementation.

Abbildung 2.1 veranschaulicht diese Bestrebungen in einem Vergleich von möglichen Zugehörigkeitsfunktionsformen welche im Bereich der Fuzzy-Logik Anwendung finden.

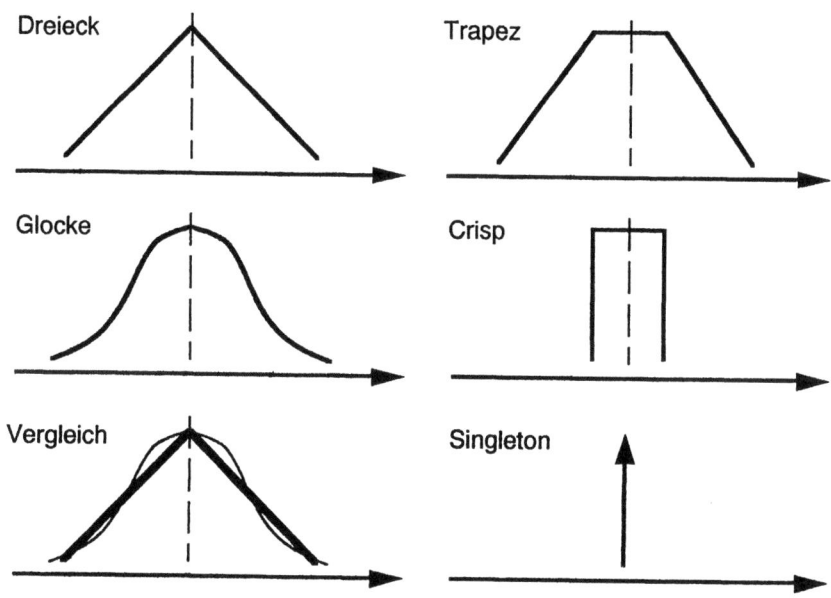

Abbildung 2.1: Zugehörigkeitsfunktionsformen

Fuzzy-Logik basiert auf dem Verständnis das gewisse Funktionen bewußt unscharf definierbar sind. Vergleicht man unter diesem Gesichtspunkt beispielsweise die geringfügige Diskrepanz zwischen einer Dreiecks- und einer Glockenfunktion, so kann man in der Regel die durch die Vereinfachung verlorengegangene Gegnauigkeit als vernachlässigbar ansehen.

2.2. Mögliche Zugehörigkeitsfunktionsformen beim NLX230

Betrachtet man eine typische Glockenfunktion als Repräsentanten einer möglichen Fuzzy-Zugehörigkeitsfunktion, so kann eine eindeutige Aussagen über das Verhältnis zwischen Eingabewert (z.B. ein von einem Sensoren-Interface übergebener Singleton Meßwert - siehe Abbildung 2.1) und Zugeghörigkeitsmaß ermittelt werden. Abbildung 2.2.1 zeigt die

Verknüpfung der Eingabewerte A und B mit den korrespondierenden Funktionszugehörigkeitsmaßen µa und µb.

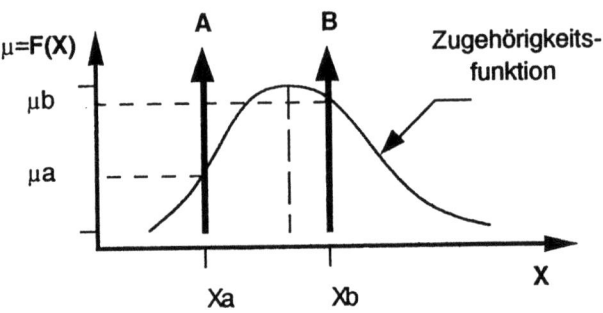

Abbildung 2.2.1: Zugehörigkeitsfunktion

Limitiert man die Formgebung der möglichen Zugehörigkeitsfunktionen im Sinne der NLX Fuzzy MicroController Philosophie, so erzielt man ein Ergebnis gemäß Abbildung 2.2.2. Die Verknüpfung von Singleton-Eingabewerten und Funktionszugehörigkeitsmaß bleibt hierbei unverändert.

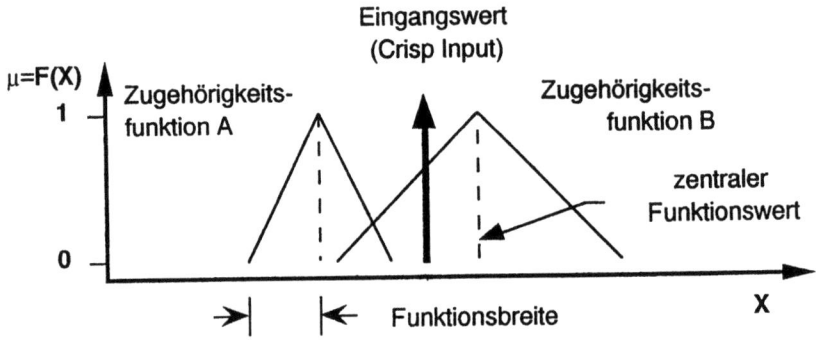

Abbildung 2.2.2: Fuzzy Zugehörigkeitsfunktionen

Reduziert man nun erneut die Zugehörigkeitsfunktionsform auf ein gleichschenkeliges und rechtwinkeliges Dreieck, so lassen sich einige interessante Merkmale feststellen. Abbildung 2.2.3 veranschaulicht die Ermittlung eines Zugehörigkeitsmaßes. Fällt ein Singleton-Eingabewert in die Funktionsbreite, d.h. das Zugehörigkeitsmaß ist >0, so läßt sich das Funktionszugehörigkeitsmaß wiefolgt darstellen:

μ_A = MAX - |(zentraler Funktionswert - X_A)|

Wobei μ_A das Maß an Funktionszugehörigkeit darstellt, MAX das maximale Maß an Funktionszugehörigkeit angibt und X_A den Singleton-Eingabewert beschreibt.

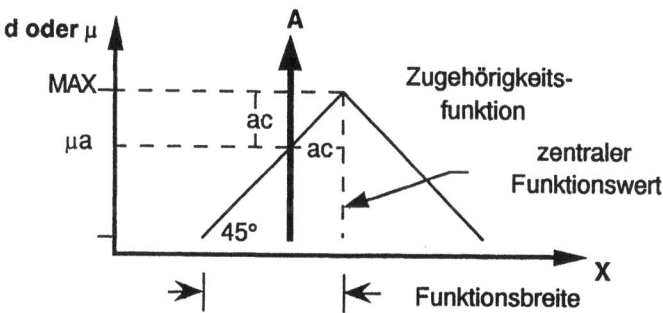

Abbildung 2.2.3: NLX230 Zugehörigkeitsfunktion

2.3 Alpha-Cut Erzeugung beim NLX230 Fuzzy MicroController

Ausgehend von konventionellen Dreieckszugehörigkeitsfunktionen gemäß Abbildung 2.3.1.1, unterscheidet man beim NLX230 Fuzzy MicroController zwischen dem in Abbildung 2.2.3 skizzierten gleichschenkeligen und rechtwinkeligen Dreieck und dem gekappten Dreieck gemäß Abbildung 2.3.1.2. Vergleicht man die Alpha-Cut Erzeugung zwischen konventionellen Dreiecksfunktionen (Abb. 2.3.1.1) und gleichschenkeligen und rechtwinkeligen Dreiecken (siehe Abb. 2.3.1.2), so sind die erzielten Ergebnisse in beiden Verfahren schlüssig. Es gilt anzumerken, daß durch das kappen der Dreiecksfunktion beim NLX230 die Erzeugung des Funktionszugehörigkeitsmaßes nicht beeinträchtigt wird. Dies wäre nicht der Fall, wenn man die Dreiecksfunktionen strecken und quetschen könnte. Um das hier erläuterte Verfahren der Erzeugung eines Zugehörigkeitsmaßes anwenden zu können müssen beide Bedingungen, d.h. Gleichschenkeligkeit und Rechtwinkeligkeit, erfüllt sein.

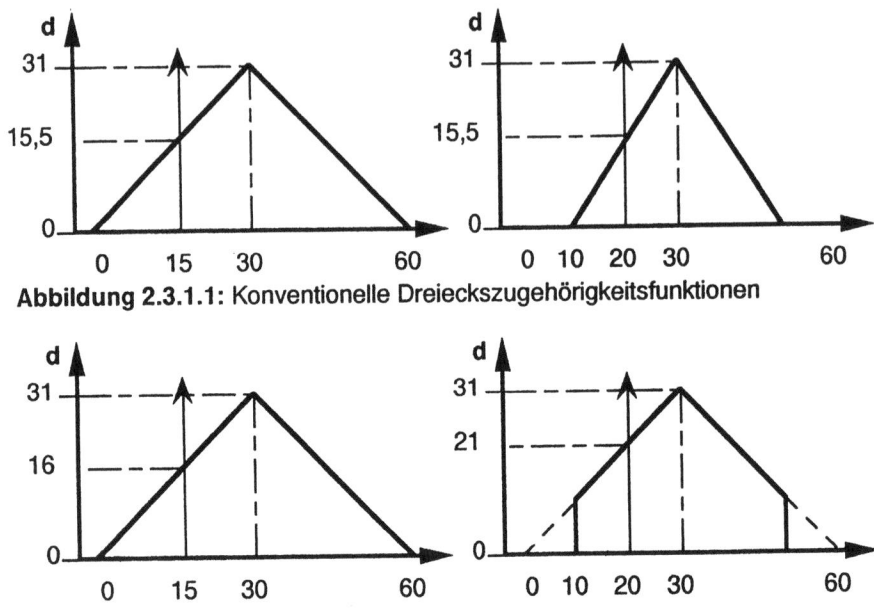

Abbildung 2.3.1.1: Konventionelle Dreieckszugehörigkeitsfunktionen

Abbildung 2.3.1.2: Fuzzy MicroController Zugehörigkeitsfunktionen

Die eigentliche Erzeugung des Zugehörigkeitsmaßes, d.h. die Errechnung des Alpha-Cut Wertes verläuft gemäß Abbildung 2.3.2.

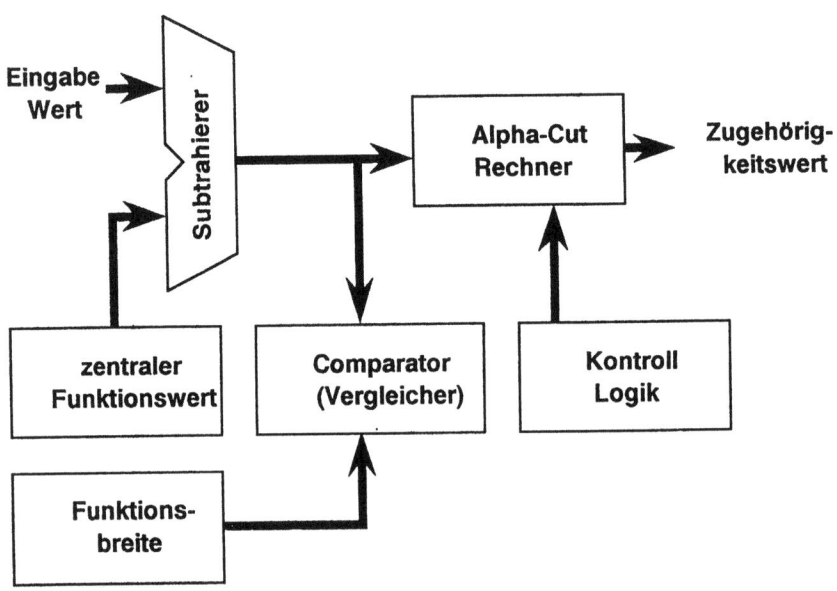

Abbildung 2.3.2: Ermittlung des Zugehörigkeitswertes

3 Prozeß-Steuerung mit Fuzzy Controller Chips

Der Fuzzy-Logik Control Regelkreislauf (Abbildung 3.1) folgt weitgehend dem konventionellen Regelungsprozeß.

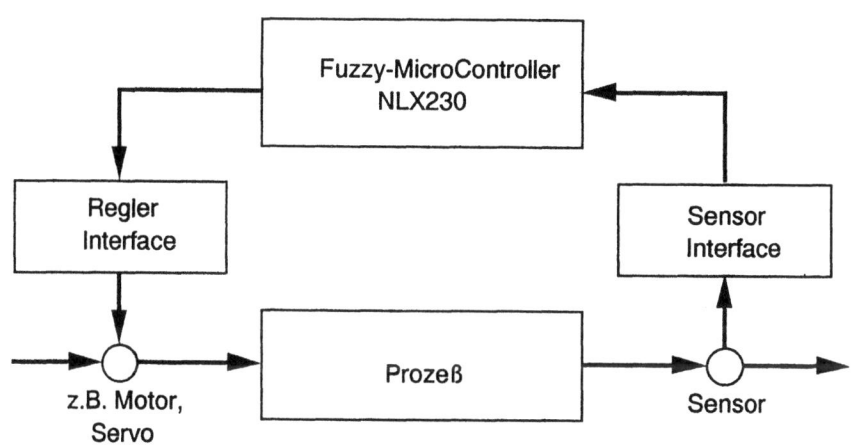

Abbildung 3.1: Fuzzy-Logik Control Regelkreislauf

Gegenüber regulären Mikroprozessoren, welche für jeden anliegenden Eingabeparameter einen sequentiell ablaufenden Algorithmus inizieren um einen modifizierten Ausgabewert zu erzeugen, verwendet der NLX230 Fuzzy MicroController einen parallelen Datendurchsatze, d.h. optimale Ausgabewerte werden kontinuierlich erzeugt. Die Verarbeitungsgeschwindigkeit beträgt hier ca. 30 Millionen Regeln pro Sekunde. Sämtliche Fuzzy-Arbeitsschritte, d.h. Fuzzifizierung, Überprüfung der Regelmatrix und Defuzzifizierung (siehe Abbildung 3.2), werden im Fuzzy MicroController parallel und kontinuierlich ausgeführt. Applikationen welche die Stärken des Fuzzy MicroController Steuerungssystems bestmöglich ausnutzen liefern präzise Ausgabeparameter mit einer bis zu 1200 mal höheren Geschwindigkeit als dies unter Zuhilfenahme eines konventionellen Mikroprozessors realisierbar war. Derartige Vergleichszahlen sind mit Sicherheit nur unter Vorbehalt relevant, da der direkte Vergleich in starkem Maß von der Applikation abhängig ist. Dennoch erscheint eine Hardware Lösung zur Fuzzy-Logik eine schlüssige Vortsetzung der Bemühungen um das Vorantreiben dieser neuen Technologie.

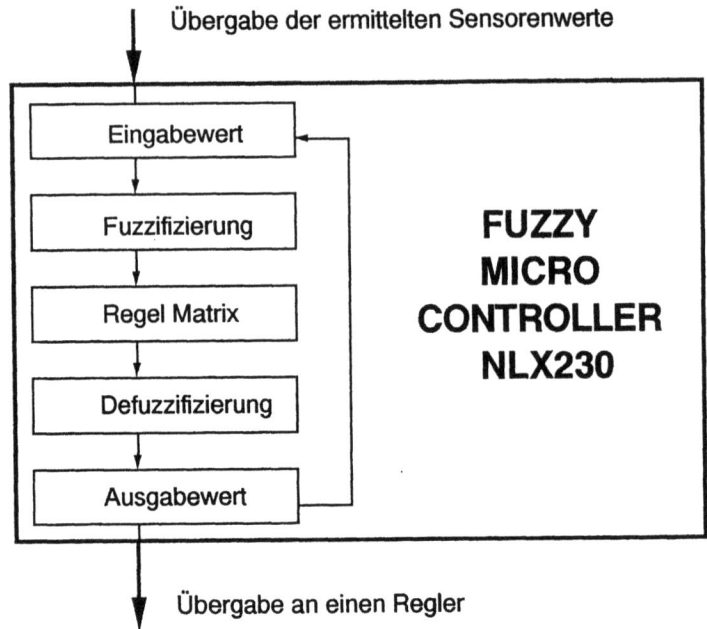

Abbildung 3.2: Möglicher Prozeßablauf in einem Fuzzy-MicroController

Der Fuzzy Controller-Chips ist in der Lage eine Auflösung von 5-Bit (Zugehörigkeitsmaß: $31 \geq \mu \geq 0$) * 8-Bit (Singleton Eingabewert Spektrum: $255 \geq \mu \geq 0$) zu unterstützen (siehe Abbildung 3.3).

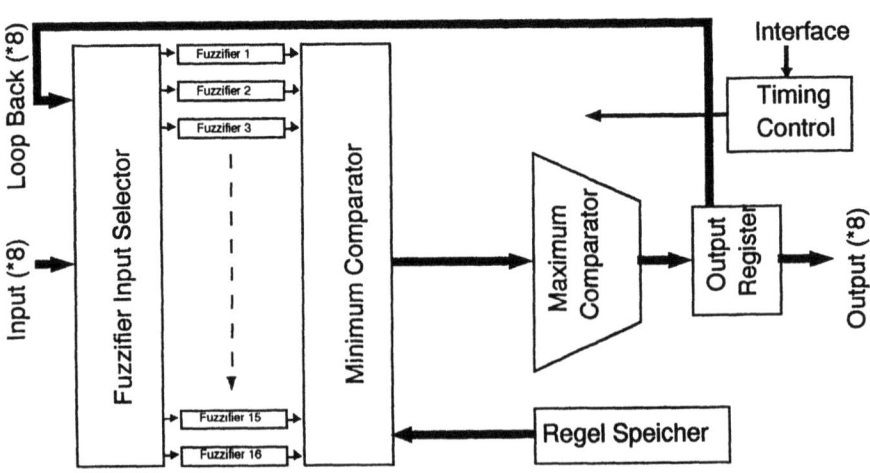

Abbildung 3.3: Blockdiagramm des NLX230 FuzzyMicrocontrollers

Der NLX230 erlaubt bis zu je acht Ein- und Ausgänge, wobei es möglich ist, durch eine Loop-back Verknüpfung Ausgänge wiederum als Eingänge zurückzuführen. Es lassen sich bis zu sechzehn frei verfügbare Fuzzifizierungsstufen (für die Festlegung der linguistischen Terme) mit den acht Eingängen verknüpfen. Die Defuzzifizierung folgt hierbei der bewährten MIN/MAX Methode.

4 Applikation einer Fuzzy Steuerung am Beispiel

Die Applikation einer Fuzzy Steuerung sei hier am Beispiel einer Staubsauger Steuerung illustriert. Bei der Entwicklung der Kontroll-Logik gilt es zunächst die sensorisch ermittelten Eingangswerte (vom Sensor-Interface übergebene Singleton Werte) für den Ansaugluftdruck, für die Schmutzstärke, und für die Oberflächenbeschaffenheit den jeweils unterschiedlichen Fuzzifizierungsstufen zuzuordnen (siehe Abbildung 4.1.1.). Das unter der Anleitung von Expertenwissen entwickelte Regelwerk leistet dann die kausale Verknüpfung gemäß den Projektanforderungen.

4.1 Die Festlegung von Linguistischen Termen

Abbildung 4.1.1 zeigt die Term Editor Definitionsmaske der NeuraLogix ADS230 Entwicklungsoberfläche. Die drei Eingabeparameter sind hier mit ihren korrespondierenden linguistischen Termstufen dargestellt. Unterscheidet man beim Luftdruck (Abbildung 4.1.2) und beim anliegenden Verschmutzungsgrad (Abbildung 4.1.3) auf fünf linguistische Terme (von sehr_klein bis sehr_groß bzw. von gering bis stark), so reichen für die Beschreibung der Oberflächenbeschaffenheit (Abbildung 4.1.4 und Abbildung 4.1.5) hier bereits drei Terme aus (glatt, mittelrauh und rauh). Insgesamt sind in dem hier vorliegenden Staubsaugerbeispiel drei Eingangswerte mit 13 Zugehörigkeitsfunktionen (linguistischen Termen) verbunden.

```
═FMC Development System - by American NeuraLogix, Inc.═
   File      Edit      Download      Configure        Run
                     ──Term Editor──

   Input              Fuzzifier         Center Width Membership
   Druck           is D_sehrklein         0     30   Inclusive
   Druck           is D_klein            40     20   Inclusive
   Druck           is D_mittel           70     20   Inclusive
   Druck           is D_gross            90     20   Inclusive
   Druck           is D_sehrgross       135     30   Inclusive
   Schmutz         is S_gering            0     30   Inclusive
   Schmutz         is S_mittelgering     40     20   Inclusive
   Schmutz         is S_mittel           70     20   Inclusive
   Schmutz         is S_mittelstark     100     20   Inclusive
   Schmutz         is S_stark           127     20   Inclusive
   Oberflaeche     is O_glatt             0     25   Inclusive
   Oberflaeche     is O_mittelrauh       30     20   Inclusive
   Oberflaeche     is O_rauh             60     30   Inclusive
                   is                     0      0   Inclusive
                   is                     0      0   Inclusive
                   is                     0      0   Inclusive


Enter an Input name      F1-Help F2-Choose F3-Graphic F10-Save
```

Abbildung 4.1.1: Termeditor des ADS230 Entwicklungssystems

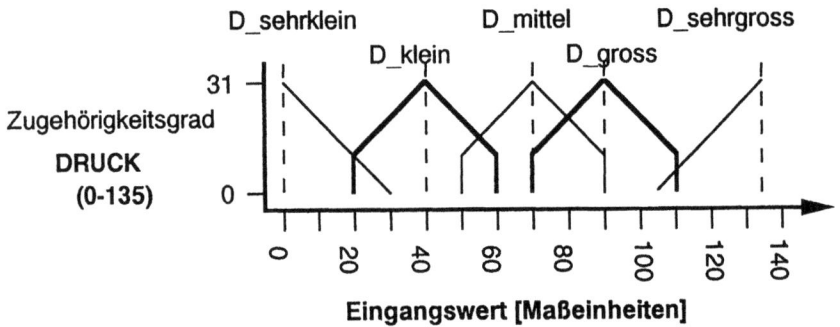

Abbildung 4.1.2: Zugehörigkeitsfunktionen des DRUCK Eingabewertes

Ermittelt die Sensorik eine Oberflächenbeschaffenheit von 20 [Maßeinheiten], so findet dieser Wert ein Zugehörigkeitsmaß von 21 (mit MIN=0 und MAX=31 bei einer 5-bit Auflösung) zu der Funktion 'mittelrauh'. Da die linguistischen Terme 'mittelrauh' und 'glatt' sich zwischen den Eingangswerten 9-24 [Maßeinheiten] überschneiden, erzielt eine Oberflächenbeschaffenheit von 20 [Maßeinheiten] außerdem ein Zugehörigkeitsmaß von 11 (mit MIN=0 und MAX=31 bei einer 5-bit

Auflösung) zur Menge 'glatt'. Mit Hilfe des graphischen Term-Editors ist es möglich die Position (Zentrum) und Deckungsbreite der einzelnen Terme beliebig zu verändern. Die Entwicklung von Fuzzy-Termen bedarf somit keinerlei Pseudo-Kode-Programmierung, sondern kann stattdessen wahlweise durch benutzerfreundliche Menusteuerung realisiert.

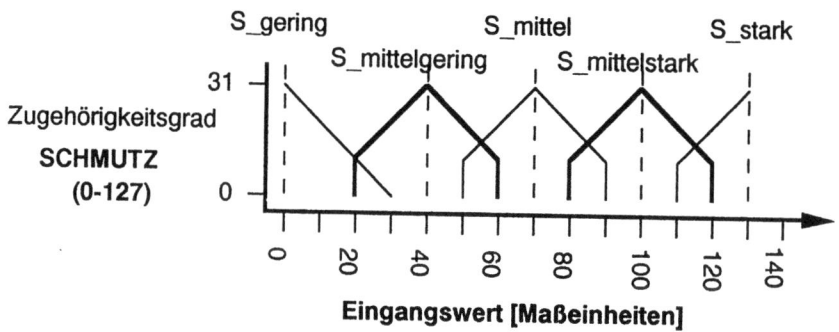

Abbildung 4.1.3: Zugehörigkeitsfunktionen Eingabewert SCHMUTZ

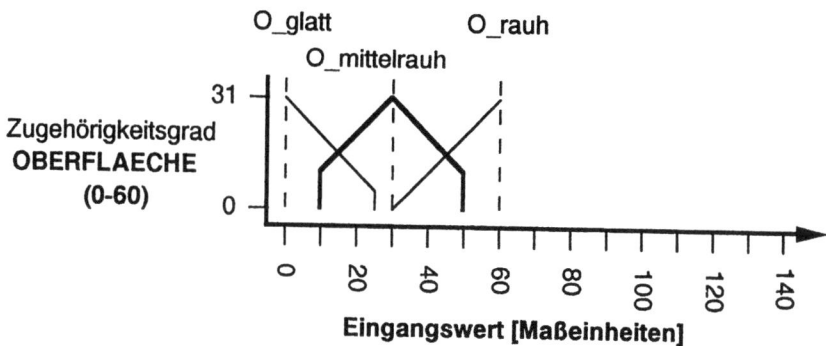

Abbildung 4.1.4: Zugehörigkeitsfunktionen des Eingabewertes OBERFLÄCHE

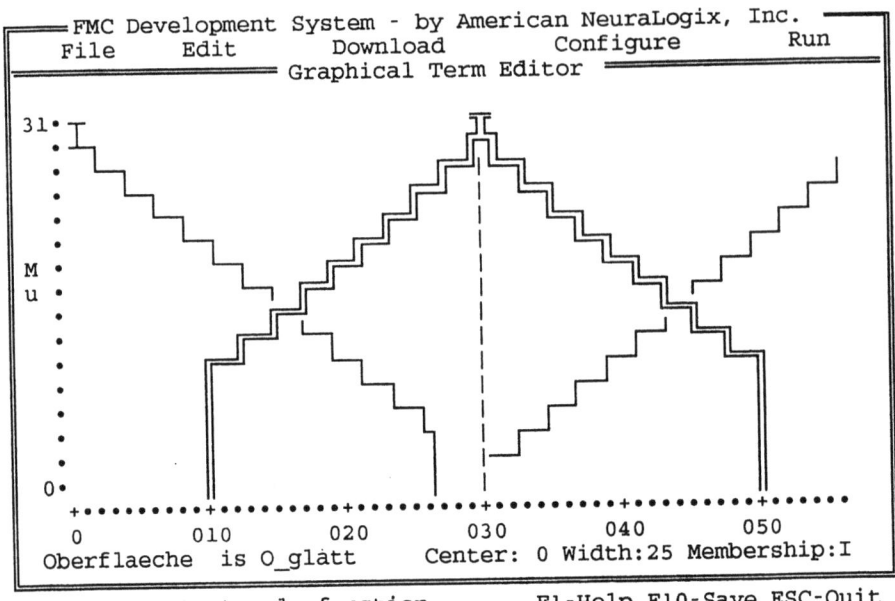

Abbildung 4.1.5: Graphischer Zugehörigkeitsfunktionseditor des NeuraLogix ADS230 Entwicklungssystems

4.2 Die Erzeugung einer Regel Matrix

Die Regelmatrix beinhaltet die logische Verknüpfung von fuzzifizierten Eingabewerten mit korrespondierenden Ausgabewerten (Sog. Action-Values). Bei der Erstellung des Regelwerkes ist ein hohes Maß an Expertenwissen und die Analyse von empirischen Fallbeispielen von großer Bedeutung. Hierzu ein Beispiel: Es gilt die drei Eingabeparameter, Ansaugluftdruck, Schmutzstärke, und Oberflächenbeschaffenheit, mit dem Ausgabewert Ansaugdruckstärke zu koppeln. Erfassen die Sensoren die Eingabeparameter 'geringer' Druck, 'starker' Schmutzgehalt, und 'rauhe' Oberflächenbeschaffenheit, so erscheint es sinnvoll die Ansaugdruckstärke zu erhöhen. Die Umsetzung des intuitiven Regelwerkes in ein rechnerverständliches Format geschieht mit der Hilfe des graphischen Regel-Editors (Abbildung 4.2.1).

```
╔══FMC Development System - by American NeuraLogix, Inc. ══╗
║   File      Edit      Download       Configure        Run ║
║                    ─── Rule Editor ───                    ║
║                        Rule   1                           ║
║  If      Druck              is  D_sehrklein               ║
║  and X   Druck              is  D_klein                   ║
║  and     Druck              is  D_mittel                  ║
║  and     Druck              is  D_gross                   ║
║  and     Druck              is  D_sehrgross               ║
║  and     Schmutz            is  S_gering                  ║
║  and     Schmutz            is  S_mittelgering            ║
║  and     Schmutz            is  S_mittel                  ║
║  and X   Schmutz            is  S_mittelstark             ║
║  and     Schmutz            is  S_stark                   ║
║  and     Oberflaeche        is  O_glatt                   ║
║  and     Oberflaeche        is  O_mittelrauh              ║
║  and X   Oberflaeche        is  O_rauh                    ║
║  and                        is                            ║
║  and                        is                            ║
║  and                        is                            ║
║  then    Luftdruck          =>  50                        ║
╚═══════════════════════════════════════════════════════════╝
  Press space to include or exclude Term      F1-Help F10-Save ESC
```

Abbildung 4.2.1:   Individuelle Regelanalyse auf dem ADS230 Entwicklungssystems

Die drei Eingangswerte (Druck, Schmutz, und Oberfläche) sind durch eine Regel-Matrix miteinander verknüpft (die Regel-Matrix ist zur rechten Seite fortgesetzt und kann durch Scrolling auf dem Bildschirm sichtbar gemacht werden). Greift man eine bestimmte Regel heraus, so wird die Verknüpfung deutlich. In Abbildung 4.2.1 erzielt Regel 1 folgende Aussage: Wenn der Druck 'gering' ist und der Schmutz 'stark' ist und die Oberfläche 'rauh' ist, dann erhöhe den Ansaugdruck (um 50 [Einheiten]). In gleichem Maße werden sämtliche vorliegenden Regeln parallel verarbeitet.

Der NLX230 erzielt den jeweils optimalen Ausgabewert unter der Zuhilfenahme des Minimum-Maximum Prinzips. Zunächst wird für jede individuelle Regel das kleinste vorliegende Zugehörigkeitsmaß ermittelt. Aus allen relevanten Regeln wird dann die Regel mit dem größten Maß an Zugehörigkeit ausgewählt und der Ausgangswert dieser Regel wird ausgeführt. Abbildung 4.2.2 verdeutlicht dieses Verfahren am Beispiel des Staubsaugers. Folgende Eigabewerte liegen an der Sensorschnittstelle an: Druck=40 [Einheiten], Schmutz=125 [Einheiten] und Oberfläche=55 [Einheiten]. Dies korrespondiert relativ deckend mit den linguistischen Termen 'geringer' Druck, 'starke' Verschmutzung und 'rauhe' Oberfläche. Intuitive erscheint es unter diesen Umständen sinnvoll den Ansaugdruck zu erhöhen. Fünf Regeln sind für die Kontrolle des Ansaugdrucks von Bedeutung (Abbildung 4.2.2).

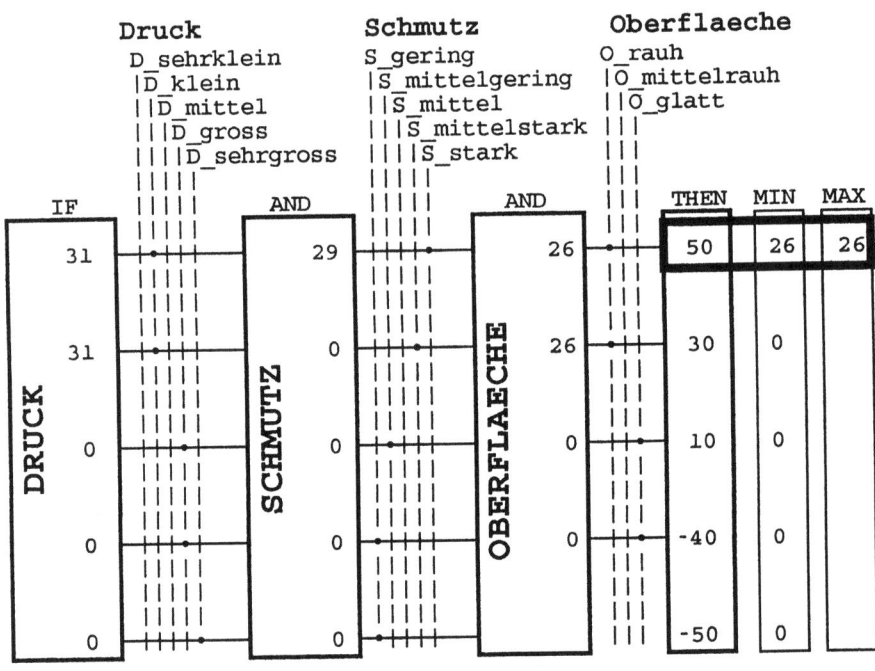

Abbildung 4.2.2: Ermittlung der siegenden Regel am Beispiel des Ansaugdrucks

Das Zugehörigkeitsmaß der drei Eingabeparameter zu den dreizehn korrespondierenden linguistischen Termen wird nun in die fünf relevanten Regeln substituiert. Demnach erfüllt die erste Regel folgendes Maß an Zugehörigkeit: 'geringer' Druck=31 (von 0=MIN bis 31=MAX), 'starker' Schmutz=29 (von 0=MIN bis 31=MAX) und 'rauhe' Oberfläche=26 (von 0=MIN bis 31=MAX). Für die erste Regel ist somit der minimalste auftretende Wert = 26. Der minimalste auftretende Wert in den verbleibenden Regeln ist in allen vier Fällen gleich null. Somit ist der maximale Wert, welcher aus den fünf relevanten Regeln hervorgeht in der ersten Regel enthalten. Dies bedeutet, daß der Ansaugdruck-Ausgabewert stark erhöht wird (+50 [Einheiten]). Diese Resultat entspricht somit den zu Beginn erläuterten intuitiven Erwartungen.

5 Schlußbemerkung

Mit dem Einzug der Fuzzy-Logik hat die Mikroelektronik mit Sicherheit eine weitere Marktniche erschlossen. Das günstige Verhältnis von Preis und Geschwindigkeit machen den Fuzzy MicroController zu einer Alternative in einer großen Zahl von konventionellen Mikro-Controller Applikationen. Insbesondere im Bereich der Haushaltsgeräte, Konsumelektronik, Automobilelektronik und Prozeßsteuerung, haben Fuzzy-Controller potentielle Chancen. Der Markt steht den neuen Geräten mit "eingebautem Intelligenzquotzienten" mit Sicherheit aufgeschlossen gegenüber.

Touchpad
Fuzzy bewertet Fingerdruck

Michael Hertzberg
RaPoTronic
Poppner & Rachel OHG
Hauert 8
4600 Dortmund 50

Einleitung

Die Produktpalette der Firma RaPoTronic, ansässig im Technologiepark Dortmund, umfaßt kundenspezifische Steuerungen auf Mikroprozessorbasis für Maschinen unterschiedlicher Art, z.B. Steuerungen für Kompressorverbundsysteme, Lagerpaternoster, Werkzeugmaschinen und Handhabungssteuerungen. Diese Steuerungen werden größtenteils komplett mit Fronttafel bzw. Bedienfeld und Kurzhubtastatur geliefert. Die verwendeten Steuerprogramme erfordern aufgrund ihrer grafischen Bedienoberflächen die Benutzung einer Maus. Allerdings ist diese den Anforderungen einer rauhen Industrieumgebung nicht gewachsen. Als alternatives Eingabemedium wurde hierzu das Touchpad entwickelt.

Das Touchpad

Das Touchpad ist ein universell einsetzbarer Mausersatz für rauhe Industrieumgebung. D.h. es ist gedacht für Umgebungen, in denen eine normale Maus aufgrund von Verschmutzungen und mechanischen Einwirkungen nicht einsetzbar ist. Es handelt sich hier um einen berührungsempfindlichen Sensor, der keine mechanischen Komponenten enthält. Somit ist er gegen mechanische Einwirkungen unempfindlich. Wird der Sensor in ein Bedienfeld integriert, so ist die Schutzart IP65 erreichbar. Somit kann er direkt an der Maschine installiert werden, Öl, Schmiermittel, Späne und andere Umwelteinflüsse können dem Touchpad nichts anhaben.

Der Sensor arbeitet nach einem resistiven Verfahren, so daß man bei Ausübung eines Druckes auf die Sensorfolie, sowohl ein Kraft- als auch ein Positionssignal bekommt. Ein nachgeschalteter Microcontroller wertet das analoge Signal aus und wandelt es in verschiedene Datenformate um, u.a. in das gängige Mausformat, so daß das Touchpad direkt als Mausersatz an die serielle Schnittstelle des PC angeschlossen werden kann.

Eigenschaften des Sensors

Der Sensor hat auch einige Nachteile. Um ein sauberes Positionssignal zu erhalten, muß ein recht kräftiger Fingerdruck ausgeübt werden, der zu Ermüdungserscheinungen führen kann.
Generell hat der Sensor folgende Eigenschaften:

- stabiles Positionssignal bei kräftigem Fingerdruck
- instabiles Positionssignal bei leichtem Fingerdruck

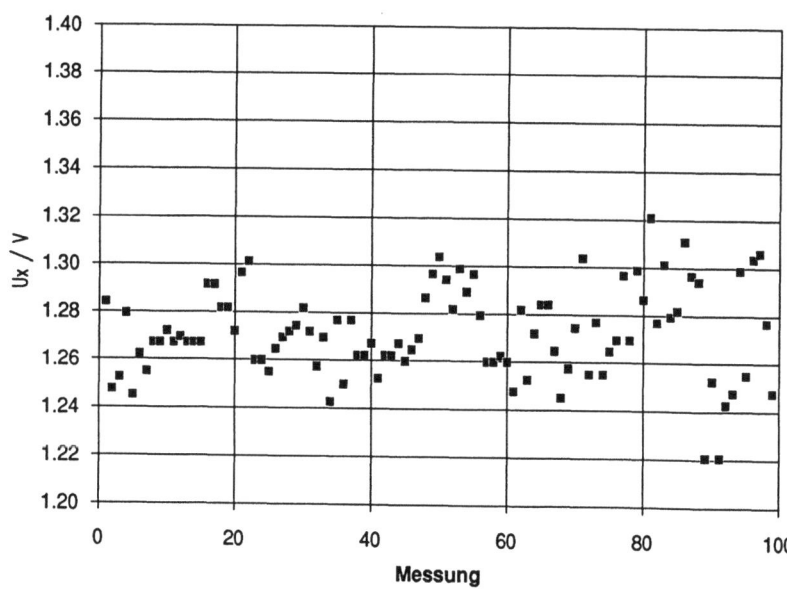

Bild 1a: Reproduzierbarkeit eines Positionssignals bei wiederholter Belastung (100g) eines Punktes

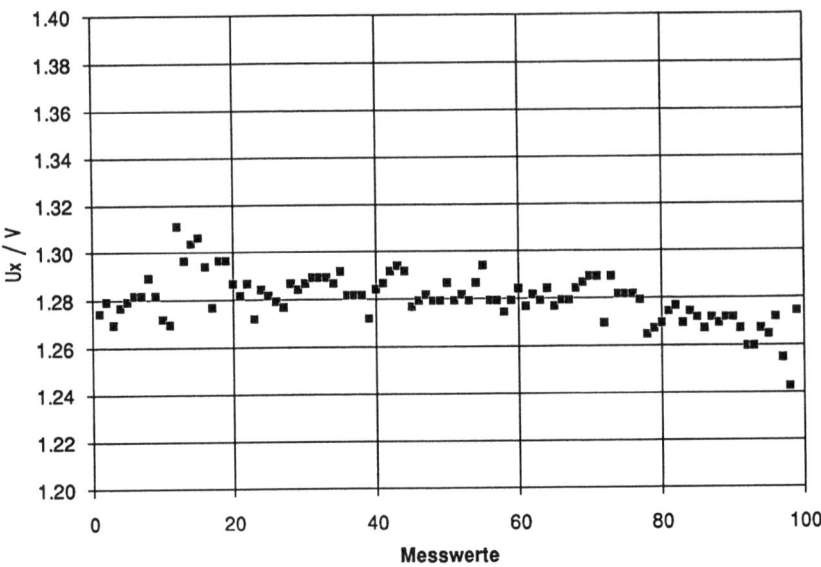

Bild 1b: Reproduzierbarkeit eines Positionssignals bei wiederholter Belastung (500g) eines Punktes

Bild 1 zeigt den Zusammenhang zwischen ausgeübter Kraft und dem gemessenen Positionssignal. Bei dieser Messung wurde eine definierte Position einmal mit einem 100g-Gewicht und einmal mit einem 500g-Gewicht belastet und das erzeugte Positionssignal gemessen. Diese Messung wurde mehrfach wiederholt. Man sieht deutlich, daß die Signalschwankung bei hoher Auflagekraft (500g) wesentlich geringer ist als bei 100g. Die ursprüngliche Touchpadversion erforderte ein Auflagegewicht von ca. 200 - 250g. Ziel ist es, diese Kraftschwelle auf ca.100g zu reduzieren.

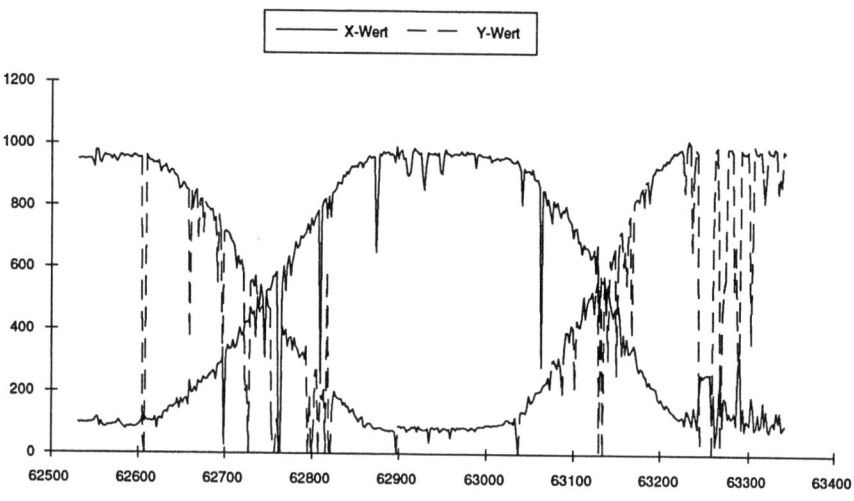

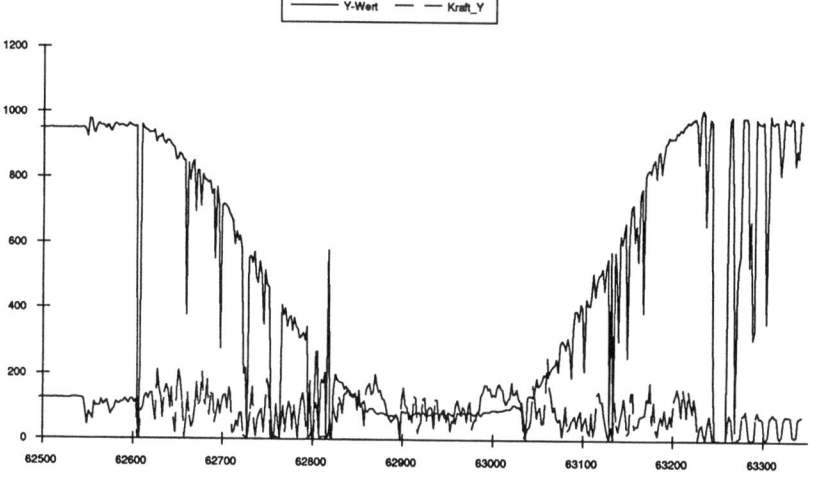

Bild 2 : gleichmäßige diagonale Bewegung eines 100g - Gewichtes

Als 2. Beispiel sehen Sie in Bild 2 eine gleichmäßige diagonale Bewegung eines 100g-Gewichtes. Anhand dieser Darstellung sieht man deutlich die extremen Meßwertschwankungen. Diese treten hauptsächlich nach unten auf, nur sehr selten nach oben! Dies geschieht immer, wenn die Kraft nachläßt.

An dieser Stelle soll nun die Fuzzy-Logic integriert werden, die das instabile Positionssignal bei 100g bewertet und gegebenenfalls korrigiert, so daß wieder ein stabiles Signal entsteht.

Ziel: - gutes Positionssignal bei leichtem Fingerdruck !

In konventioneller C-Technik ist eine Meßwertkorrektur sehr aufwendig und unübersichtlich, da sehr viele einzelne Abfragen und Vergleiche durchgeführt werden müssen. Deshalb wird die Fuzzy-Technologie eingesetzt, von der erwartet wird, daß diese den Entwicklungsprozeß des erforderlichen C-Codes erheblich beschleunigt und überschaubar macht. Mit Fuzzy ist eine "flexible Beurteilung" möglich, durch den Einsatz linguistischer Variablen nähert sich die Problemstellung der menschlichen Denkweise und wird nicht sofort in mathematisch exakte Denkprozesse gezwängt.

Fuzzy-Implementierung

Zunächst werden die Erfahrungen der Entwickler in "wenn-dann"-Regeln beschrieben. In unserem Fall betrachtet man den Zusammenhang zwischen Auflagekraft und Positionsänderung.

Ein Beispiel:
Ist die Positionsänderung klein und der Druck hoch, dann ist das Vertrauen in den Meßwert hoch.
Ist die Positionsänderung hoch und der Druck klein, dann ist das Vertrauen in den Meßwert gering.

Für die Werte der linguistischen Variablen werden die Membershipfunktionen definiert. Die Festlegung der Werte erfolgt nach dem Erfahrungswert und dem persönlichen Eindruck des Entwicklers.

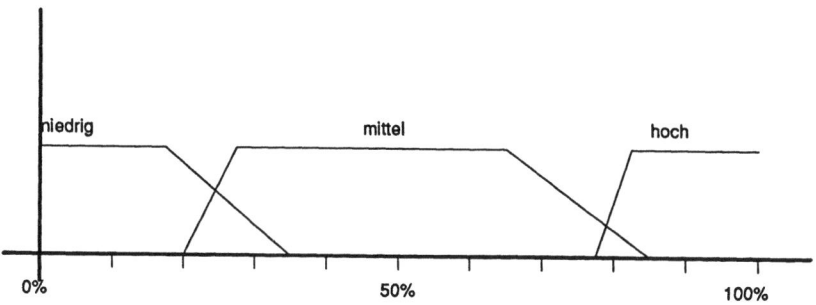

Bild 3 : Membershipfunktionen

Für jede linguistische Variable wird eine Funktion aufgestellt, eine für den Druck und eine für die Positionsdifferenz.

Der Fuzzy-Compiler generiert aus den aufgestellten Regeln einen C-Code, der in das bestehende C-Programm eingebunden werden kann. Dieser C-Code beinhaltet allerdings fast ausschließlich floating-point-Operationen, die für einen Einsatz in einem Microcontroller zu zeitaufwendig sind. Deshalb wurde der generierte C-Code in Integer-Arithmetik konvertiert.

Der entstehende C-Code kann nun in das vorhandene Touchpad-C-Programm integriert werden. Es wird als externe Funktion eingebunden und als Funktion aufgerufen. Als Eingabeparameter werden die aktuellen Meßwerte, also x-Position, y-Position und die beiden Kraft-Signale der einzelnen Sensorfolien übergeben. Die Funktion verändert gegebenenfalls die Positionswerte, wie oben beschrieben. Diese Werte stehen dann dem ursprünglichen Programm wieder zur weiteren Verarbeitung zur Verfügung.

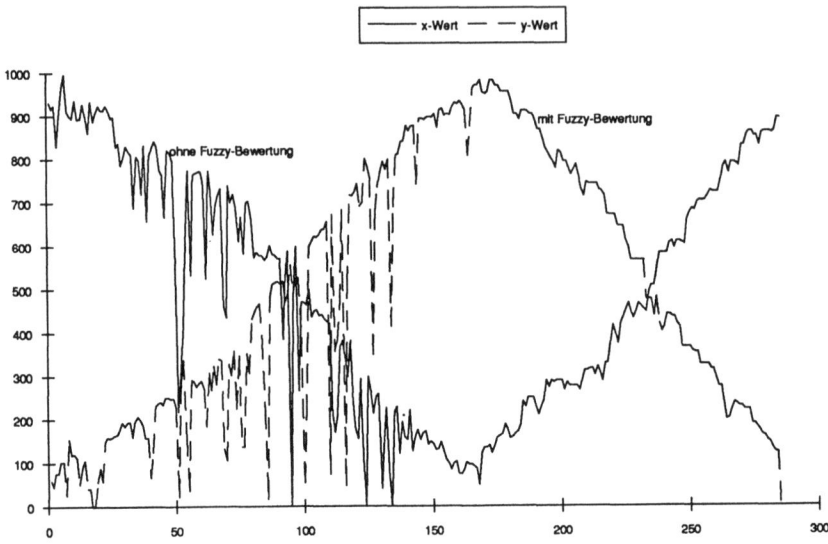

Bild 4 : erste Ergebnisse

Die ersten Ergebnisse zeigen eine deutliche Verbesserung der Meßwerte. In Bild 4 ist wieder eine gleichmäßige diagonale Bewegung dargestellt. In der linken Hälfte wird die Fuzzy-Bewertungsroutine im C-Code nicht angesprungen, es handelt sich also um die tatsächlich gemessenen Werte. Für die Zurückbewegung wurde die Fuzzy-Bewertungsroutine aktiviert, die Positionssignale sind erheblich verbessert.

Diese Ergebnisse entstammen dem ersten Programmdurchlauf. Hierbei wurden die Membershipfunktionen aus bestehenden Meßreihen aufgestellt. Eine weitere Anpassung und Optimierung dieser Funktionen verspricht eine weitere Verbesserung der Meßwerte.

Wir danken für die freundliche Unterstützung:
Fuzzy-Demonstrationszentrum Dortmund, Hr. Thomas Kretzberg

MIX
Papier aus verantwortungsvollen Quellen
Paper from responsible sources
FSC® C105338

If you have any concerns about our products,
you can contact us on
ProductSafety@springernature.com

In case Publisher is established outside the EU,
the EU authorized representative is:
**Springer Nature Customer Service Center GmbH
Europaplatz 3, 69115 Heidelberg, Germany**

Printed by Libri Plureos GmbH
in Hamburg, Germany